Tomasz Szlagor
Leszek A. Wieliczko

Vought F4U Corsair

vol. I

KAGERO

MORE FROM KAGERO

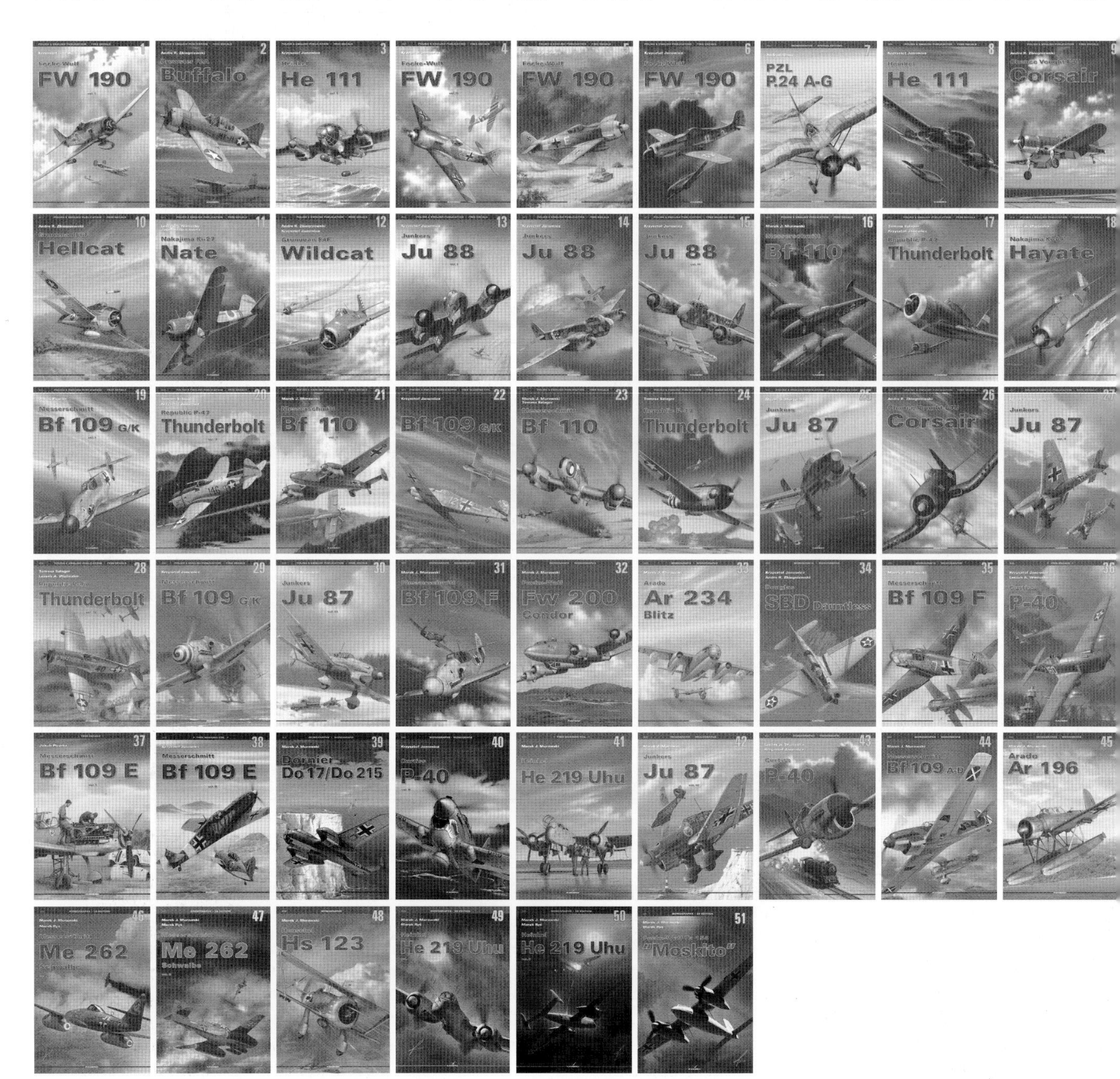

www.kagero.pl • phone +48 81 501 21 05

Vought F4U Corsair Vol. I • Tomasz Szlagor, Leszek A. Wieliczko • First edition • LUBLIN 2013

 • ISBN 978-83-62878-73-4

Editing: **Leszek A. Wieliczko** • Translation: **Tomasz Szlagor** • Cover artwork: **Arkadiusz Wróbel** • Color artwork: **Zbigniew Kolacha** • Photo credits: **Goodyear, NARA, NASA, NNAM, SDASM, US DoD, US Navy, Vought** • Design: **KAGERO STUDIO, Marcin Wachowicz**

Oficyna Wydawnicza KAGERO
Akacjowa 100, Turka, os. Borek, 20-258 Lublin 62, Poland, phone/fax: (+48) 81 501 21 05
www.kagero.pl • e-mail: kagero@kagero.pl, marketing@kagero.pl
w w w . k a g e r o . p l
Distribution: **KAGERO Publishing Sp. z o.o.**

XF4U-1 (BuNo 1443) prototype during one of test flights. The aircraft was first flown on 29th May 1940 from Bridgeport Municipal Airport by Vought's chief test pilot, Lyman A. Bullard. The prototype was painted with silver aluminum dope, only upper wing surfaces were glossy Orange (Chrome) Yellow.

The origin and development

Vought F4U Corsair was one of the most successful and renowned fighter aircraft of World War II. Developed since 1938, it didn't make it to serial production until mid-1942. Initially disqualified by the US Navy for carrier service, it was handed over to land-based US Marine Corps fighter units, where it soon proved its worth in combat. It was fast, robust and heavily armed, which also made it an excellent weapon against ground targets. When it was finally qualified for carrier operations, before long it became (along with F6F Hellcat) the primary aircraft of US Navy fighter and fighter-bomber squadrons.

The Concept

In early 1935 BuAer (Bureau of Aeronautics) of the US Navy issued a request for proposals for a new carrier-based fighter aircraft, a future successor of biplane fighters Boeing F4B and Grumman FF, operated by the Navy at that time, as well as Grumman F2F and F3F, which were shortly to enter service. The Navy accepted XF2A-1, the project submitted by Brewster company. The prototype, powered by Wright R-1820 Cyclone engine, was first flown in December 1937. BuAer also ordered – as an alternative, in case Brewster's project failed to meet expectations – Grumman's new fighter powered by Pratt & Whitney R-1830 Twin Wasp engine, which soon evolved from biplane XF4F-1 to monoplane XF4F-2. The prototype XF4F-2 was first flown in September 1937, even earlier than its competitor, but problems with its powerplant slowed down its development. Hence, F2A-1 Buffalo entered service as the first US monoplane carrier-based fighter. Eventually however, it was the Grumman fighter, the improved F4F-3 model later known as Wildcat, which proved a much more capable aircraft and in early forties became the standard US Navy carrier fighter.

Meanwhile, on 1st February 1938, a few months after first flights of XF4F-2 and XF2A-1, BuAer offered a tender for yet another carrier fighter. The new aircraft was expected to reach at least 350 mph (563 kph) maximum speed at 20,000 feet (6096 meters), have stall speed no higher than 70 mph (113 kph) and range of 1000 miles (1609 km). It was to be armed with four machine guns. As was the case three years earlier, BuAer didn't insist on any particular powerplant or design concept, besides the obvious requirement that the aircraft was ca-

XF4U-1 was the biggest and heaviest US single-engine fighter to date. It owned its distinctive look to the inverted gull wing design, earning the aircraft the name of 'Bent-Winged Bird'.

XF4U-1 was armed with two fuselage-mounted .30 inch (7.62 mm) machine guns (with muzzles in the engine cowl ring) and two wing-mounted .50 inch (12.7 mm) guns. Wing roots housed air inlets for oil coolers, intercooler and supercharger.

pable of operating from carrier decks. In fact, for the Navy the priority was the aircraft's high performance, especially maximum speed. BuAer was so focused on this factor that it became a standing joke among Vought designers, who claimed that the Navy gave them only three requirements: firstly – speed, secondly – speed, thirdly – more speed.

In response to this latest tender, in April 1938 Chance Vought Aircraft Division of United Aircraft Corporation, based at East Hartford, Connecticut, submitted two projects of a classic, single-engined, single-seat fighter: V-166A (known to BuAer as Vought A), powered by a proven R-1830 Twin Wasp engine, and V-166B (Vought B) powered by a new XR-2800 Double

Wasp engine constructed by Pratt & Whitney, still under development at that time. The latter powerplant was a huge, two-row, 18-cylinder radial engine with a displacement of 2800 cubic inches (45.9 l). The prototype XR-2800-2 (B-series) was equipped with two-stage, two-speed supercharger with intercooler. Its maximum rated power on takeoff was 1,850 hp at 2,600 rpm, and it could develop 1,500 hp of continuous power at 2,400 rpm (at 17,500 ft). In comparison, the XR-1830-76 engine powering XF4F-3 prototype produced only 1,200 hp on takeoff and 1,000 hp at 19,000 ft.

The designers from Pratt & Whitney expected soon to increase the takeoff power to 2000 hp. The R-2800 was the first American 18-cylinder engine and the first producing such enormous power for its time. The decision to mount it in a new aircraft design was a risky one, for it was a completely new and unproven powerplant; on the other hand, there was a good chance that the Navy would get its desired high-performance fighter.[1]

BuAer, having analyzed all submitted proposals, selected the one referred to as the Model V-166B and powered by XR-2800-2 engine. On 11th (or 30th, according to other sources) June 1938 Vought was awarded a contract No. 61544 authorizing the company to build a prototype of the new fighter, which was given military designation XF4U-1 and serial number (BuAer Number, BuNo) 1443. The same month BuAer signed a similar contract with Grumman for building a prototype of XF5F-1, a twin-engined fighter, and in November also with Bell for building XFL-1 prototype powered by Allison V-1710 inline engine. Of these three very novel fighter aircraft designs, only Vought XF4U-1 made it to serial production.

XF4U-1 viewed from below. Lower fuselage featured an oval window, and there were two bomb bays in each wing (with five cells each), where national markings were applied, for a total of forty 5.20 lb (2.36 kg) anti-aircraft bombs.

The Design

The team of designers working on technical details of XF4U-1 included Frank C. Albright (as project engineer, replaced in January 1941 by John Russell “Russ” Clark); Paul S. Baker and William C. Schoolfield (as aerodynamics engineers); James Shoemaker and Donald J. Jordan (propulsion engineers). The overall supervision of the project was given to Rex Buran Beisel, the company’s chief engineer. Beisel also had a decisive voice in choosing particular design solutions.

The aircraft design was conform to the requirement of achieving the greatest possible maximum speed. The key to success was ob-

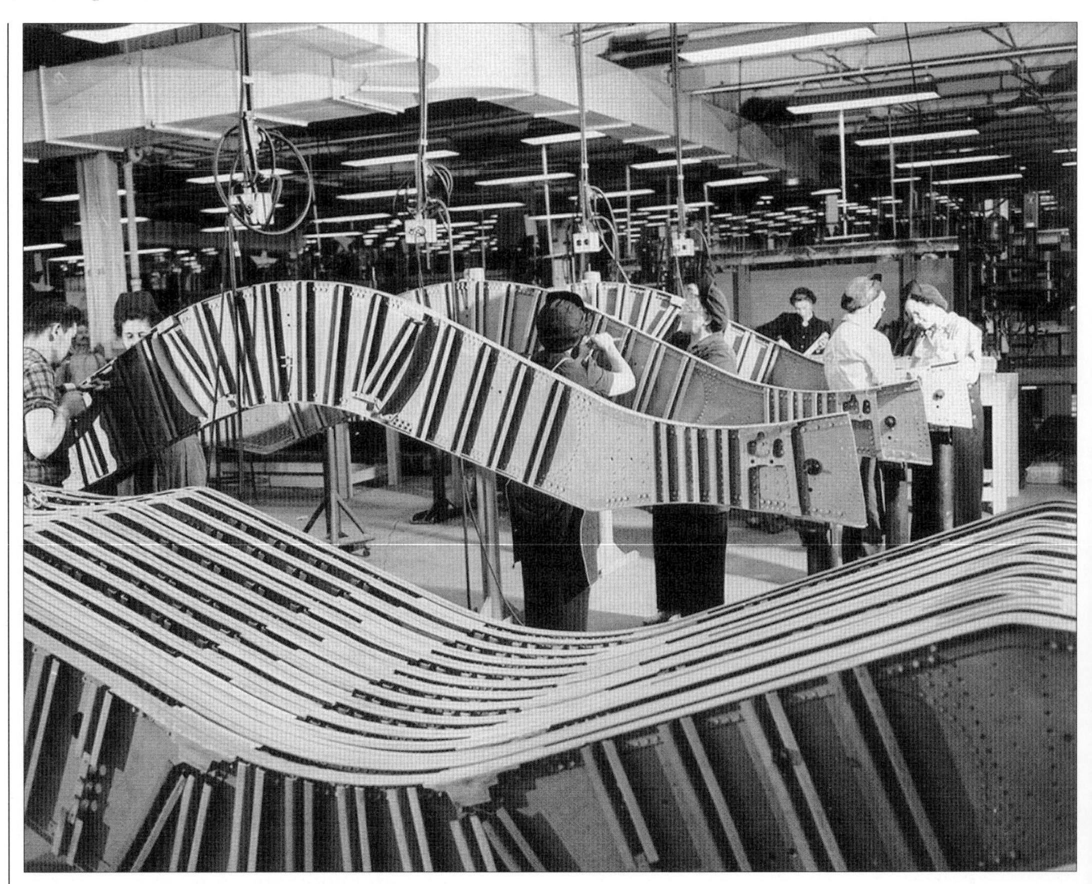

Preassembling main spars in Goodyear plant at Akron. The spars were the main construction element of the aircraft, connecting both halves of the wing center section to each other and to the fuselage. Their complex shape was the result of the 'bent' wing design.

viously the R-2800 engine and careful aerodynamic refinement of the airframe. Double Wasp was 52.8 in (1,342 mm) in diameter – a little less than R-1820 (54.25 in; 1,378 mm) but significantly more than R-1830 (48.03 in; 1,220 mm). However, it was almost twice as heavy and much longer, which meant that it had to be coupled to a fittingly big and sturdy (thus heavy) airframe. In fact, XF4U-1 was the largest and heaviest of all American single-engined fighters built to date. Its empty weight was about 7500 lbs (3,400 kg) – about 2700 lbs (1,200 kg) more than XF4F-3 and over 3300 lbs (1,500 kg) more than XF2A-1.

Port wing flaps in the lowered position. A retractable spanner was mounted between the inner and center flap sections. When the flaps were lowered, the spanner covered the gap between the two sections.

In order to harness the engine's tremendous power, a large-diameter propeller was necessary. Hamilton Standard company (which, like Pratt & Whitney, was a division of United Aircraft Corp.) came up with a purpose-designed, three-bladed constant-speed propeller of 13 foot 4 inch (4.06 meter) in diameter – the largest of all hitherto used in single-engined aircraft. Now the question was, how to achieve sufficient propeller clearance. Longer landing gear struts were not an option, because it would be very difficult (if not impossible) to fit them inside wings, and reinforced struts would increase the aircraft's weight. The designers came up with a very innovative solution – the inverted gull wing, with wings bent down from the root at the angle of 23 degrees, then canted gently upwards (the dihedral at the upward bend was 8.5 degrees).[2] The main landing gear was attached at the lowest point of the 'crank' in the wing, so the struts could be comparatively short and light.

Such wing design had another very welcome feature. The wings were attached to the fuselage at a nearly perfect right angle, which was an optimal solution as far as aerodynamics went, as it helped minimize interference drag – there was no need for mounting wing-root fairings (overlaying intersections between wings and fuselage). Wing roots, which were the thickest part of the wing, housed oil coolers, with air inlets at the wing leading edges and vent doors at the bottom surface of the wings. The same inlets supplied air for both stages of the supercharger (depending on the engine operating condition, one or both stages could be employed) and intercooler.[3] The air outlet of the intercooler was located under the fuselage. All these features allowed the designers to retain a clean airframe and further reduce drag. The only disadvantage of such wing layout was a complex structure of the main spar (connecting both halves of the wing center section to each other and to the fuselage), which also had to have the shape of a flattened W letter when looked from the front.

The wings were of single-spar construction, with an auxiliary rear spar in the wing center section. They consisted of rectangular-shaped center section and two tapering outer panels with rounded tips. The outer panels could be folded up over the canopy for maintenance and storage at carrier decks. The aircraft's height with folded wings was 16 ft 4 in (4.98 m), and its width 17 ft 0.61 in (5.19 m). In order to accommodate oil coolers and main landing gear

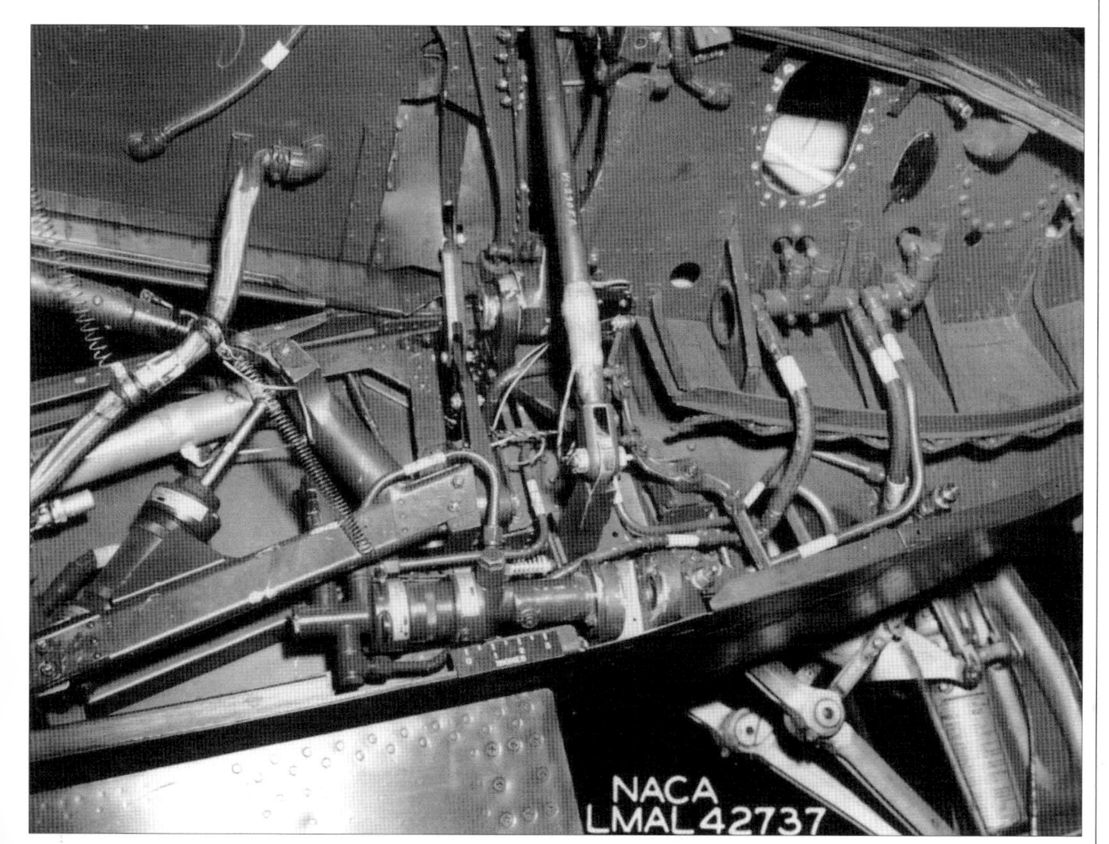

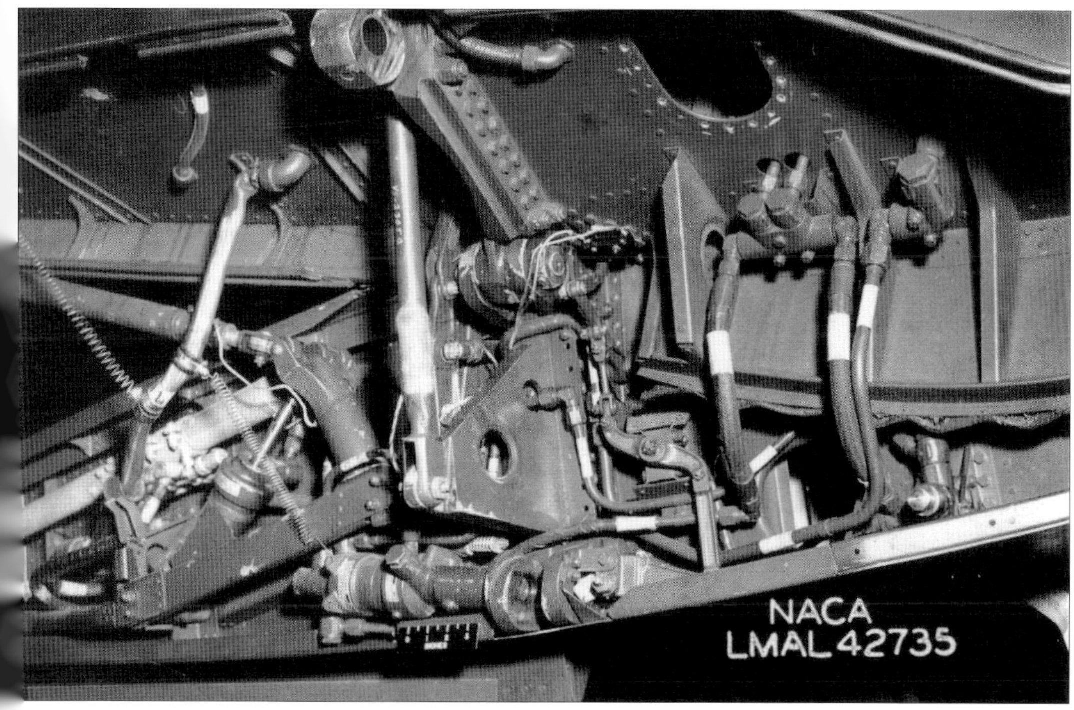

Starboard wing fold mechanism. Like many other mechanisms of the aircraft, including landing gear and flaps, it was hydraulically operated.

XF4U-1 was demolished in a crash that occurred on 11th July 1940, and rebuilt in two months' time. On 1st October it reached the top speed of 405 mph (652 kph), becoming the first US single-engine fighter to exceed 400 mph (644 kph) in level flight.

XF4U-1 prototype at the factory airfield's compass calibration pad, in April 1941. The aircraft has a new propeller with broader blades and more rounded tips. Retractable landing light can be seen under port wing.

wells in the wing center section, the designers chose a fairly thick NACA 23000 series airfoil, which thickness ratio changed smoothly from 18% at the roots to 15% at the junction with outer panels, and 9% at the wingtips. Main landing gear retracted rearward, with the wheels rotating through a 90 degree to lay flat in a fully enclosed wheel well. Wheel dimensions were 32×8 in (813×203 mm), and the wheel track was 12 ft 1 in (3.68 m). Interestingly, the main landing gear, especially the flat strut covers, could act as aerodynamic brakes reducing the aircraft's diving speed. A pilot could select to lower the main landing gear only, without lowering the tail wheel and arresting hook. However, in practice this option was rarely used.

Wings were equipped with ailerons and flaps, the latter subdivided into three separate

XF4U-1 tested at NACA, Langley, in autumn 1941. Of note is a long instrument probe mounted under starboard wing. The aircraft's very smooth skinning was achieved by extensive use of spot welding.

sections. The three-section arrangement was necessary because of the bent shape of the wing and the folding of the outer wing panels. Two flap sections were carried by the wing center section, and one by the outer wing panel. The flaps were of a new type, known as slotted deflector flaps, designed and patented by Roger W. Griswold, and used for the first time in OS2U Kingfisher observation-scout floatplane.

The flaps were fitted with deflector plates attached to the flaps' leading edges. When the flaps were being lowered, the plates shielded from above the slot between the flaps and wings, deflecting the airflow (hence their name). In this way greater angles could be obtained without stalling the flap and having the air break away from the upper flap surface and thereby cause the flap to lose its lift. The new flaps were much more effective than the standard slotted flaps, although their construction was practically the same.

The flaps were all metal except for outer sections, which were fabric-covered. Ailerons had wood frames and plywood skinning; the port aileron was fitted with a trim tab. Outer wing panels had metal construction and fabric skinning aft of the main spar, except for an area inboard under the gun bay, which was metal covered.

The size of the fuselage was determined by the engine diameter. Immediately aft of the engine the fuselage gradually tapered towards the tail, changing the cross-section from round to oval. Technologically, the fuselage was divided into front section, mid section and aft section. The front section housed a spacious cockpit with a fixed windshield and a backwards sliding canopy streamlined into the fuselage's outline. Although the cockpit was located near the wings' trailing edges, it offered a fairly good field of view ahead and downward on either side of the nose because of the bend in the wings. The cockpit had no floor, only small footrests directly in front of the rudder pedals. The bottom of the fuselage featured a small teardrop shaped window allowing the pilot to see directly beneath the aircraft.

The front part of the fuselage was attached to the wing center section at the main spar. The engine bearer, made of welded steel tubes, was attached to the front part of the fuselage. The engine had a closely fitting cowling, with 18 cowl flaps completely encircling the fuselage aft of the engine. The engine and its accessory compartment was covered with detachable panels, which offered easy access to the engine, supercharger and other installations.

Mid section accommodated radio set and other equipment. The rear section of the fuselage housed the tail wheel bay covered with doors, and the rearmost tip was used for mounting a retractable arrester hook. Tail planes were tapered, with rounded tips. Small fin, to which a broad-chord rudder was attached, was located ahead of the horizontal stabilizers. Rudder and elevators were all fitted with trim tabs, and elevators additionally with balance tabs.

While constructing the fuselage, the designers made extensive use of spot welding – an innovative method developed together with engineers of Naval Aircraft Factory in Philadelphia and used for the first time also in OS2U floatplane. Instead of a dense 'skeleton' of longerons and formers, with small panels of skinning riveted to it, a light framework was used, consisting of only several main bulkheads and four main longerons (two lower and two upper ones). Large duralumin sheets (the largest of these sheets measured 48×102 in; 122×259 cm) had stiffeners spot-welded to the inboard sides. Each of these sheets is preshaped by stretching over forms, and some of them incorporate

One of the first serial-production F4U-1s in flight, in late summer 1942. Compared to the prototype, it has the cockpit moved aft, modified canopy and window cutouts directly aft of the cockpit. Lowered landing light and Mk 41-2 bomb rack can be seen to advantage.

compound curvatures. The preassembled sheets were then flush-riveted to the framework. Such skinning was much more rigid and smoother, and it didn't crease during riveting. Furthermore, this method produced a lighter and more durable fuselage than in case of the classic, fully riveted semi-monocoque construction.

The aircraft's armament comprised two fuselage mounted Browning M2 .30 inch (7.62 mm) machine guns firing through propeller arc, with muzzles in the upper engine cowl ring, and two Browning M2 .50 inch (12.7 mm) machine guns, one in each outer wing panel. The ammunition boxes for the fuselage mounted guns had a capacity of 750 rounds, while the box for each wing mounted gun had a capacity of 300 rounds. Additionally, there were two small bomb bays with five cells each for small 5.20 lb (2.36 kg) fragmentation bombs, 40 in all, inside outer wing panels. The idea was to use these bomblets for breaking enemy bomber formations. The upper sides of the outer wing panels had built-in compartments for inflatable air bags, designed to help the aircraft remain afloat in the event of a ditching at sea, and give the pilot time to safely get out of the cockpit and retrieve a life-raft stowed in a recess behind the pilot's headrest.

The fuel was contained in four wing tanks (two in wing center section and two in outer panels), 273 US gallons (1,033 l) in all. To keep the Corsair aerodynamically as clean as possible, the designers made no provision for external auxiliary drop tanks. Hydraulic installation was used for a variety of tasks: lowering and retracting landing gear, flaps and arrester hook, opening and closing landing gear bay doors, folding and unfolding outer wing panels, operating cowl flaps, flow splitters inside air inlets, vent doors and wheel brakes, as well as recharging guns.

XF4U-1 prototype

In January 1939 the subdivisions of United Aircraft Corporations underwent reorganization. Chance Vought Aircraft and Sikorsky Aircraft were merged under the name of Vought-Sikorsky Aircraft Division of United Aircraft Corp." Sikorsky's plant at Stratford, Connecticut, became the main center of aircraft production and was duly expanded. Vought had to move from East Hartford to a new headquarters, where de

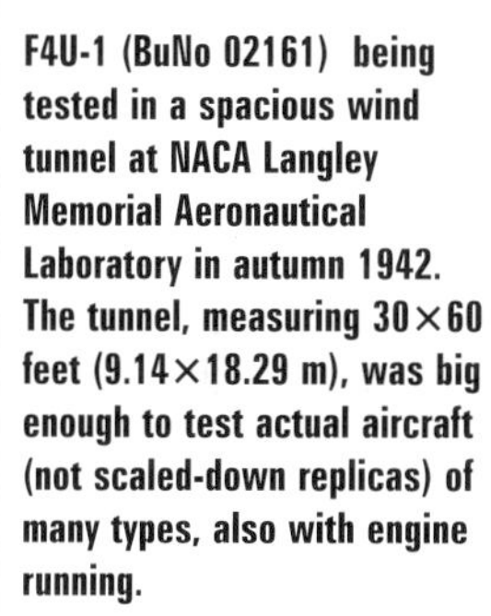

F4U-1 (BuNo 02161) being tested in a spacious wind tunnel at NACA Langley Memorial Aeronautical Laboratory in autumn 1942. The tunnel, measuring 30×60 feet (9.14×18.29 m), was big enough to test actual aircraft (not scaled-down replicas) of many types, also with engine running.

The same F4U-1 (BuNo 02161) tested for aerodynamic drag. The propeller was dismounted, engine air inlet was shielded with a streamlined fairing, and all seams were duct-taped to minimize drag.

sign work on XF4U-1 was continued. Meanwhile, on 8th-10th February 1939 a committee of Navy specialists inspected a full scale wooden mock-up of the aircraft. The designers were advised to introduce several modifications (among other things, cockpit canopy was enlarged, at the cost of adding yet another bracing frame[5]; also, Mk 3 telescopic gunsight was replaced by a more modern Mk 8 reflector gunsight[6]), whereupon BuAer accepted XF4U-1. In early July a detailed technical documentation was almost ready, and construction of a prototype commenced.

The XF4U-1 prototype (BuNo 1443) was ready in May 1940, over two years after the initial order had been placed. Once all assemblies had been checked and rechecked, ground

A factory fresh F4U-1 (BuNo 02172) being readied for a maiden flight in late summer 1942. Note the Mk 41-2 rack for 100 lb (45.4 kg) bombs. The aircraft is in the finish typical of that period: Blue Gray on upper and side surfaces and Light Gray on the undersides.

engine test runs and taxiing tests were carried out for several days. Finally, on 29th May, Lyman A. Bullard Jr, Vought's chief test pilot, took off from Bridgeport Municipal Airport[7] to test-fly the prototype. The first flight lasted 38 minutes and went smoothly, although the pilot noticed elevator flutter and some minor malfunctions of trim tabs. The following test flights posed no problems until 11th July, when during the fifth flight storms prevented test pilot Boone T. Guyton from reaching the airfield at Stratford. Running low on fuel, he attempted to land at a golf course near Norwich. On wet grass the aircraft went into a skid and plowed into trees, flipping onto its back. The outer panel of the starboard wing was torn off, empennage and propeller were smashed, but the fuselage, engine and undercarriage suffered only minor damage, which testified to the toughness of the construction. Guyton walked away practically unscathed.

The crash could have had serious consequences for the aircraft's future. Fortunately, Vought managed to rebuild the prototype in just two months. It was at this time that the XF4U-1 was fitted with an improved version of XR-2800-4 engine (power rating remained unchanged) and a new, more effective propeller of the same diameter, but with slightly broader blades and more rounded tips.

On 1st October 1940 the rebuilt prototype, with Lyman Bullard at the controls, was presented in the air to an official delegation of US Navy headed by Vice Admiral John H. Towers, chief of BuAer. During the flight from Stratford to Hartford XF4U-1 reached the top speed of 405 mph (652 kph), thus becoming the first US single-engine fighter to surpass the 'magic' barrier of 400 mph (644 kph) in level flight and proving the concept.[8]

On 24th October XF4U-1 was handed over to NAS (Naval Air Station) at Anacostia. During subsequent flight tests, which continued until the end of February 1941, the prototype demonstrated good handling and high performance for an aircraft of its size and weight, by far surpassing BuAer expectations. Nonetheless, the aircraft had its share of deficiencies – engine overheating, bad spin recovery, slow aileron response and a slight lateral instability – but these seemed easy to remedy. In mid-June the aircraft was sent to NACA at Langley Field, Virginia, for wind tunnel tests.[9]

In August simulated carrier landings were carried out at Naval Aircraft Factory in Philadelphia (with a carrier deck outlined on the runway). The tests revealed that the aircraft, when slowing down to stalling speed, had a dangerous tendency to suddenly drop port wing. At the end of August the prototype was returned to Vought, but for more than two years it continued to serve as a test-bed, both at NACA and in US Navy test facilities. It was eventually

F4U-1 assembly line at Vought plant in Stratford. In 1942 the company delivered only 178 aircraft, but before long the production went into high gear and in 1943 Vought delivered exactly ten times as many Corsairs.

struck off the Navy inventory on 22nd December 1943.

As early as 3rd March 1941, BuAer issued a preliminary proposal to Vought to start a serial production of the new fighter. Once the construction details were agreed upon, on 30th June a contract No. 82811 was signed for production and delivery of first batch of 584 F4U-1s (BuNo 02153–02736). The aircraft was officially named Corsair. The Navy expected first deliveries in February 1942. Several months later, due to the mounting tensions in the Far East, BuAer resolved to increase the number of expected F4U-1s by ordering license production at Brewster Aeronautical Corporation from Long Island City, New York, and Goodyear Aircraft Company from Akron, Ohio. The contracts were signed in November and December 1941, respectively. Brewster's Corsairs were designated F3A-1, and Goodyear's FG-1.

In March 1943 several F4U-1s were handed over to Pratt & Whitney for engine tests. One of them, seen here to the left, designated F4U-1M or F4U-1WM (BuNo 02460), was coupled to the powerful, 28-cylinder XR-4360 Wasp Major engine. Seen to the right is a standard production F4U-1.

F4U-1A armed with a 1,000 lb (454 kg) bomb mounted on centerline rack designed by Brewster company. The rack became Corsairs' standard equipment beginning with BuNo 17930. The most prominent feature of the F4U-1A is the bulged, 'blown' cockpit canopy with only two bracing frames.

F4U-1A in tricolor camouflage introduced in early 1943. The undersides of foldable wing panels were painted Intermediate Blue, just like fuselage sides and tailfin. Note the rectangular window in the lower fuselage, deleted in the course of production run. There was provision for mounting an auxiliary fuel tank of 170 gal (644 l) or 175 gal (662 l) capacity under fuselage.

F4U-1

Serial-production F4U-1 (factory designation VS-317) incorporated many modifications and improvements. They resulted not only from flight and wind tunnel tests, but also from changing requirements by BuAer, which was learning from experiences of air combat over Europe. By then the Navy wanted, besides high performance, stronger armament and more protection for the pilot and fuel tanks.

The most extensively modified part of the aircraft was its fuselage. The unprotected wing center section fuel tanks were removed, and instead a large, 237 US gal (897 l) fuel tank was installed between cockpit and engine. The new tank was protected by a thick layer of self-sealing rubber weighting 177 lbs (80 kg). In order to fit in the new tank, the designers had to lengthen fuselage by 17 in (43 cm), and move cockpit 32 in (81 cm) rearward from its original location. Cockpit canopy was again redesigned by reducing the number of bracing frames. The top frame of the windshield was fitted with Brownscope wide-angle rear-view periscope system enabling observation of the rear hemisphere. Small oval window cutouts were inset into the fuselage directly aft of the cockpit, providing the pilot with a limited rear view over his shoulders. The periscope system was soon discarded in favor of a rear-view mirror located inside a blister at the forward top panel of the canopy. Also the fuselage window cutouts proved redundant, and in some late-production machines which already had them the cutouts were faired over with metal sheets. The window in the bottom of the fuselage was moved aft along with the cockpit and it was enlarged. Relocating cockpit to the rear further impaired forward visibility in a three-point attitude, which proved especially troublesome during arrested landings.

A 1.5 inch (38 mm) pane of laminated bullet-proof glass was added behind the windshield. Steel armor plates, weighting a total of 135 lbs (61 kg), were inserted ahead of oil tank, ahead of cockpit and behind pilot's seat. Upper panels of fuselage skinning covering fuel tank and instrument panel were made of reinforced, 0.102 in (2.6 mm) duralumin sheets weighting 25 lbs (11,3 kg). Mechanisms of tailwheel and arrester hook, separate in XF4U-1, now were combined into one unit, retracted into the same bay in the rear fuselage and covered with doors. In accordance with increasingly more frequent use of radar on US Navy ships, the aircraft was equipped with IFF (Identification Friend or Foe) AN/APX-1 device.[10]

Wings underwent a major modification, too. The .30 inch guns were by then considered of little use, hence initially it was planned to strengthen the aircraft's armament by adding a second pair of wing-mounted .50 inch guns. Eventually the .30 inch guns were discarded altogether (in fact, after installing the fuselage fuel tank there was no room for them anymore), and the number of .50 inch guns was increased to six – three in outer part of each wing. The amount of ammunition was radically increased. Each of the inboard and center pair of guns was fed from two interconnected ammunition boxes of 200 rounds (which meant 400 rounds

F4U-1D in flight with two bombs on wingroot pylons. This modification turned the Corsair into a capable fighter-bomber. The F4U-1D model lacked leading edge wing tanks and upper cowl flaps.

Late-production F4U-1D (probably BuNo 82721) at Vought plant in 1945. The aircraft sports the new frameless canopy. Beginning with BuNo 82253, the Corsair could carry Mk 5-2 zero-length rocket launchers, mounted four under each outer wing panel, for 5-in (127 mm) FFAR or HVAR rockets.

per gun), whilst the guns of the outboard pair drew ammunition from boxes of 200 and 175 rounds (which gave 375 rounds per gun). The total amount of ammunition was 2,350 rounds. In order to fit in the guns and ammo boxes, the designers discarded the floatation bags (future showed that even without them a ditched aircraft stayed afloat long enough) and the wing bomb bays (the bomblets were deemed useless against bombers), but also downsized the leading edge fuel tanks, mounted outboard of the gun bays. Now each of these tanks could hold up to 63 US gal (238.5 l) of fuel. Since they were unprotected, pilots were supposed to purge them with CO_2 (supplied from a separate bottle) to get rid of volatile fumes. Total amount of fuel aboard the F4U-1 was 363 US gallons (1.374 l). As before, no provision was made to install external drop tanks. However, a single Mk 41-2 rack could be mounted under each outer wing panel for carrying light, 100 lb (45.4 kg) bombs.

Based on the results obtained from wind tunnel tests, ailerons were increased in span by 20 inches (50.8 cm), with a consequent reduction in outboard flap section span. Bigger ailerons much improved the aircraft's roll control and increased its roll rate. During production run each aileron was fitted with a balance tab. The slotted deflector flaps used in the prototype were deemed too complicated and were replaced by standard slotted flaps, which were lighter and easier to produce. The flaps' maximum deflection angle for landing was reduced from 60 to 50 degrees. Upper wing surfaces near wingtips were equipped with blue formation lights, a white recognition light was added to the upper surface of the starboard wing, and three colored (red, green and amber – front to rear) identification lights to the port wing undersurface.

Due to all these modifications, empty weight of a serial-production F4U-1 increased by over 1325 lbs (600 kg) as compared to the prototype, and takeoff weight by over 2650 lbs (1,200 kg). Nevertheless, the aircraft's performance, instead of degrading, much improved. It was possible thanks to installing the latest version of R-2800-8 engine producing 2,000 hp on takeoff. The engine drove three-bladed, constant-speed Hamilton Standard Hydromatic propeller of 13 ft 4 in (4.06 m) in diameter (with blades of 6443A-21 or 6525A-21 type). The maximum speed of the F4U-1 in clean condition increased to 417 mph (671 kph) at 20,000 feet (6,096 meters).

Since so many of the aircraft's subassemblies had to be redesigned, the start of the production was delayed. The first serial-production F4U-1 (BuNo 02153) was test-flown by Boone Guyton on 25th June 1942. On 21st July it flew to NAS Anacostia for tests. In mid-August the seventh production machine (BuNo 02159)[11] was delivered to NAS New York at Floyd Bennett Field in Brooklyn. On 25th September the same aircraft, flown by US Navy pilot LtCdr Sam Porter, was used for carrier qualifications aboard the escort carrier USS Sangamon (CVE-26) in Cheasapeake Bay. Unfortunately, the carrier landings revealed many deficiencies. The most alarming problem was the premature and sudden port wing stall at approach speed; if the pilot didn't react in time, the aircraft was very likely to flip overboard or crash against the deck. Undercarriage oleo struts were found to have bad rebound characteristics on landing, allowing the aircraft to bounce out of control down the carrier deck. Forward visibility when coming down to land in nose-up position was next to none. Worse still, valve push rods and cowl flap actuators leaked oil and hydraulic fluid, which filmed over the windshield. All these problems made the Navy disqualify the Corsair as a carrier-capable fighter, at least until the producer remedied them. For the time being the aircraft was to operate exclusively from land bases. Thereby, US Marine Corps became the main operator of early Corsairs.

The hydraulic fluid leaks seemed easiest to fix. Originally each cowl flap was operated by its own hydraulic actuator. As a temporary solution, actuators of three upper flaps were removed, and the flaps permanently fixed and sealed. Subsequent tests showed that the temperature of the upper cylinder heads did not rise significantly. In December 1942 the cowl flaps actuating system was redesigned; individual actuators were replaced by one central unit operating the flaps by means of cables and springs. Just in case, the three flaps across the top of the fuselage were left in the closed position; later, during the production run of F4U-1D, they were replaced by a fixed metal sheet panel. Additionally, the skin panel seams above the fuselage fuel tank in front of the cockpit were often covered with sealing tape. These white lines of tapes became a distinctive feature of many Corsairs.

By the end of 1942 the plant at Stratford delivered only 178 Corsairs to the Navy. The following year the production sped up, and more orders were received. Throughout 1943 Vought alone delivered 1,780 aircraft, including 758 of the early F4U-1 variant. Of the latter number, 70 machines (BuNo 18122–18191) were unofficially designated F4U-1B (B for 'British') and supplied under Lend-Lease provisions to the British, who named them Corsair Mk I and assigned identification numbers JT100–JT169.

A number of early-production Corsairs were used for various tests conducted by the producer, at US Navy facilities and at NACA. In 1943 one (BuNo 02296) was made available to US Army Air Force test center at Wright Field near Dayton, Ohio, where it was put through paces in comparative tests versus Army's Lockheed P-38G, Republic P-47C and North American P-51. Several aircraft were delivered to Pratt & Whitney for engine tests. At least one of them (BuNo 02460), provided in March 1943, was coupled to a new powerful, 28-cylinder, four-row XR-4360 Wasp Major radial engine rated at 3,000 hp and driving a four-bladed propeller.[12] The aircraft, temporarily designated F4U-1M or F4U-1WM (WM standing for 'Wasp Major', factory designation VS-336), was first flown on 12th September 1943, becoming the predecessor of F2G Super Corsair fighters produced later by Goodyear.

F4U-1A

As the production progressed, the designers strove to fix the aircraft's remaining deficiencies. In May 1943, 405th serial-production F4U-

F4U-1D (BuNo 82326) in typical late-war finish – overall glossy Sea Blue ANA623, with antiglare panel ahead of the windshield in non-specular Sea Blue ANA607, and black propeller hub. The last three digits of the serial number, '326', painted on the cowl with white washable paint, were applied only when ferrying the aircraft to the unit.

Corsair's underside. Two 150 US gal (568 l) fuel tanks are rigged to wingroot pylons. Of note are attachment points for the catapult bridle and, further aft, rectangular window.

1 (BuNo 02557) received a new cockpit hood. Now the windshield and the rearward-sliding canopy each had only two supporting frames, and the canopy was fitted with a domed top. The pilot's seat was raised by 7 inches (18 cm), and the range of vertical adjustment was increased to 9 inches (23 cm). The new semi-bubble canopy offered a much improved all-around visibility. It was introduced beginning with 689th serial-production aircraft (BuNo 17456).[13] The F4U-1A designation, used to distinguish the aircraft with the new hood from the older 'birdcage' types, was never officially approved of by the Navy. In documentation the newer Corsairs were still referred to as the F4U-1 and such designation was stenciled on aircraft's rudders.

The biggest success was solving the problem of the port wing drop on landing. The solution proved exceptionally simple – a small (1 inch [2.5 cm] wide and 6 inches [15.2 cm] long) wooden spoiler was added to the leading edge of the starboard wing, just outboard of machine gun ports. At high angles of attack the spoiler disturbed the airflow over the starboard wing, reducing its lift. Now, as the aircraft slowed down on landing approach, both wings stalled at the same time, keeping the aircraft level. The spoiler was first tested on BuNo 02510, and introduced to serial production beginning with 943rd machine (BuNo 17710).

The problem of bouncy shock absorbers was solved after nearly a year of intense testing. Redesigned oil valves and increased air pressure made the struts stiffer, and they no longer rebounded so much during hard deck landings. BuNo 50080 introduced tailwheel strut lengthened by 6.48 inches (16.5 cm), which decreased the aircraft's ground angle (the angle between the aircraft's longitudinal axis and the ground) from 13.5 to 11.5 degrees. This modification helped improve visibility over the nose when taxiing on the ground. Furthermore, the arrester hook's maximum deflection angle was reduced from 75 to 65 degrees, which also made carrier landings easier. Beforehand, with the larger deflection angle, the hook used to hit the deck first and frequently bounced off, missing the arresting cables.

Some operational squadrons experimented with makeshift underbelly bomb racks made of steel tubes, for mounting a single 1,000 lb (454 kg) bomb. When the idea was proved in practice, Brewster designed a similar, special rack of the same lifting capacity. The rack became Corsair's standard equipment starting from 1163rd aircraft (BuNo 17930). Optionally, an auxiliary fuel tank of 170 gal (644 l) or 175 gal (662 l) capacity could be mounted instead. At the same time bomb shackles under outer wing panels, which proved of little use, were discarded. BuNo 55910 introduced the R-2800-8W engine fitted with water-methanol injection system, which increased the war emergency power to 2,250 hp at sea level. The boost installation was supplied from three tanks of overall 10.3 gal

Port wingroot pylon, here covered with a streamlined fairing. The flap in front of it is the oil cooler air outlet. The aircraft lacks intercooler flap, hence elements of engine installations can be seen.

Unguided 5-inch (127 mm) FFAR rockets, mounted under Corsairs' wings, were used during the battle for Okinawa in spring 1945. The missiles were fired from Mk 5-2 zero-length rocket launchers, which only slightly increased drag.

F4U-1D fuselage upper decking, between engine and fuel tank, stripped of skinning. Seen from bottom to top are: oil tank (with the filling point on the right side, directly ahead antenna mast), cylindrical-shaped hydraulic fluid tank and water/methanol tank with air intake ducts supplying air to supercharger.

(39 l) capacity, giving maximum duration of 8.5 minutes, provided that single continuous use didn't exceed 5 minutes. The new engine was first tested in BuNo 02625 in November 1943.

Several other modifications were introduced during production run of F4U-1A. The bottom cockpit window, deemed impractical, was deleted (the window was painted over on many of the aircraft which had it). Ammunition capacity for the outboard pair of guns was increased to 400 rounds (and the total amount of ammunition onboard to 2,400 rounds). Beginning with BuNo 17930, a landing light, previously lowered from port wing undersurface, was relocated to port wing leading edge, between undercarriage and machine guns.

Most of the above-mentioned modifications were, if possible, retrofitted to earlier machines during repairs and overhauls. This process was facilitated by the fact that as early as November 1942 Vought and the Navy created a dedicated service center at NAS North Island in San Diego, California, which overhauled and modified Corsairs serving in the Pacific theater, so that the company's plant at Stratford didn't have to get involved.

By spring 1944 Vought manufactured 2,056 Corsairs of F4U-1A model. Twenty five aircraft, most if not all of them with the old 'birdcage' canopies, were taken over by Great Britain as Corsair Mk I (BuNo 17592–17616, JT170–JT194). Another batch of 360 aircraft, picked up from different production blocks, but all featuring the new cockpit hoods, were received by the British as Corsair Mk II (JT195–JT554). The basic difference, typical of all subsequent British Corsairs, was wings shortened by 8 inches (20.3 cm) (the wingspan was reduced to 12.09 m). Clipping the wingtips – as suggested by LtCdr R.M. Smeeton of the Royal Navy, the assistant naval attaché in Washington D.C. and member of the British Liaison Office – was to facilitate folded stowage in the low-headroom hangars of British carriers. This modification, accepted in July 1943, was first tested in Corsair Mk II JT270 (BuNo 17952). Clipped wingtips had little effect on the aircraft's performance – stall speed and takeoff run marginally increased, but maneuverability at low altitude improved. The earlier delivered Corsair Mk IIs were accordingly modified, and in all the subsequent aircraft for the British clipped wingtips were done at the plant.

F4U-1D

Corsair's success as a fighter-bomber led to further development in this area in early 1944, all the more because the aircraft could possibly carry much heavier loads. Two streamlined pylons were mounted under wing center section, one on each inner wing near the wingroot, between fuselage and landing gear wells. Each of the two pylons could carry a 1,000 lb (454 kg) bomb or an auxiliary 150 US gal (568 l) drop tank. The two pylons could carry two bombs, two fuel tanks or a bomb and a tank apiece. When the pylons were not in use, teardrop shaped fairings covered their bottom to reduce drag. The F4U-1D retained the centerline hardpoint for a bomb or a drop tank, which much increased the number of possible configurations of external stores.

The two wingroot pylons, first tested on F4U-1A BuNo 50280, were incorporated into the serial production on BuNo 50350 and subsequent. The latest model was designated, this time officially, F4U-1D. The aircraft retained all the previous modifications successively introduced during the production run of F4U-1A. The leading edge wing tanks were eliminated. The armor plate ahead of oil tank was removed and the armor behind pilot's seat strengthened. The overall weight of armor plating increased to 190 lbs (86 kg).

BuNo 57356 introduced a new Hamilton Standard propeller of reduced diameter (13 ft 1 in [3.99 m], as opposed to 13 ft 4 in [4.06 m] used before), with blades of 6501A-0 or 6541A-0 type. Beginning with BuNo 57484, white recognition light was removed from starboard wing upper surface. BuNo 57583 was the first Corsair with a clear-vision 'bubble' canopy (the two overhead frames were eliminated). BuNo 82253 introduced a further enhancement of Corsair's air-to-ground (and air-to-sea) attack capabilities in the form of Mk 5-2 zero-length rocket launchers, mounted four under each outer wing section, for 5-in (127 mm) Forward Firing Aircraft Rockets (FFAR) or High Velocity Aircraft Rockets (HVAR).[14]

Before F4U-1D added the rockets to Corsair's standard weaponry, tests with rockets had been carried out on several F4U-1As. The wingroot pylons could also carry powerful 11.75 inch (300 mm) 'Tiny Tim' rockets, which nevertheless saw a very limited use during WWII. Because of rocket launchers mounted under wings, outer flap sections, hitherto fabric-skinned, had to be made of duralumin. Another modification introduced during the production run was a cut-out step in starboard inboard flap section.

F4U-1D was the first Corsair model to serve in numbers aboard US Navy carriers. By February 1945 Vought delivered 1,685 F4U-1Ds (the construction of the last two, BuNo 82853 and 82854, was cancelled), of which 150 were delivered to the British, who named them Corsair Mk II (identification numbers JT555–JT704). Altogether, Vought built 4,700 F4U-1s of all versions and variants (including the prototype and

F4U-1C (BuNo 82277) awaiting assignment. This model was being manufactured concurrently with F4U-1D. The only external difference was the four 20 mm Hispano M2 cannons in F4U-1C instead of six .50 inch machine guns in F4U-1D.

The same F4U-1C (BuNo 82277) at the factory airfield. The aircraft was fitted with the smaller-diameter (13 ft 1 in; 3.99 m) propeller. Curiously, the aircraft sports the tricolor camouflage, despite the fact that all 200 aircraft of this model were built in the latter part of 1944, when the overall ANA623 Glossy Sea Blue painting scheme was used.

F4U-1Cs described below), and together with Brewster and Goodyear – 9,442 aircraft.

F4U-1C

The F4U-1C model (C for 'Cannon-armed') came into being in July 1944 in response to BuAer's much earlier request for a more dedicated ground-attack aircraft. The six .50 in machine guns were replaced by four 20 mm Hispano M2 cannons, with a total of 924 rounds in store. This armament was first tested on BuNo 02154 in August 1943. Briggs Manufacturing company, one of Vought's subcontractors, produced 200 sets of cannon-armed wings, with deliveries to Stratford beginning in mid-June 1944. The F4U-1Cs were assembled concurrently with F4U-1Ds, and the former's serial numbers were taken from the number block of the latter.

Apart from changed armament and some minor modifications of hydraulic installation and engine accessories, F4U-1C was no different from late-production F4U-1D. Both shared the same frameless cockpit hood and propeller with smaller diameter. F4U-1C only rarely carried underwing pylons and rocket launchers, which were standard in F4U-1D. Although a 20 mm round had more hitting power than a single .50 slug, the F4U-1C was not very popular among pilots, who preferred the six 'fifties' with enough ammunition for about 30 seconds of continuous fire.[15] Moreover, unguided rockets proved more effective than cannon fire against ground targets. In the event, only 200 F4U-1Cs were built.

F3A-1

The first F3A-1 (BuNo 04515) was flown by Brewster test pilot Woodward Burke on 26th April 1943. Deliveries started in June, but initially they were small and irregular, with only 136 aircraft delivered to the Navy by the year's end. Some parts and subassemblies were produced in Long Island and Newark plants, and the final assembly was carried out in Johnsville, Pennsylvania. Brewster's Corsairs had terrible reputation – they were plagued with malfunctions, and the quality of manufacturing left much to be desired. Besides, the company was notoriously behind schedule. For these reasons, in June 1944 BuAer terminated the contract with Brewster, which by that time had produced only 735 Corsairs. Orders for the remaining 733 aircraft were cancelled. The Navy used their share of F3A-1s mainly for training. As many as 430 were delivered to the British, who designated them Corsair Mk III (identification numbers JS469–JS888 and JT963–JT972).

Initially F3A-1s were equivalent to early F4U-1s. During the production run F3A-1 incorporated modifications made to F4U-1A: new canopy with bulged top panel and two bracing frames, longer tailwheel leg, ventral hardpoint, landing light repositioned to port wing leading edge (from BuNo 04592) and R-2800-8W engine (from BuNo 11208). As in case of F4U-1A, the F3A-1A designation was never formally approved of by BuAer and wasn't even used in practice. Modifications typical of F4U-1D model were not introduced by the time the production of F3A-1 ceased. Of interest is the fact that Brewster suggested a name of Battler for the Corsairs they produced, but BuAer would have none of it.

FG-1/FG-1A/FG-1D

The first FG-1 (BuNo 12992) produced by Goodyear was flown on 25th February 1943. Deliveries to the Navy commenced in April, reaching 377 aircraft by the year's end. Goodyear proved a trustworthy associate contractor – it manufactured 2,108 Corsairs in 1944, and 1,522 more by September 1945. All in all, Goodyear's plant at Akron produced 4,007 Corsairs. Orders for remaining 755 aircraft of FG-1D model were cancelled by the Navy when the war came to an end.

Much like F3A-1, the FG-1 corresponded to F4U-1 with 'birdcage' canopy, and successively incorporated modifications introduced to F4U-1A. It's not known exactly when the new cockpit hood and taller tailwheel strut were implemented. BuNo 13261 introduced the landing light relocated to port wing leading edge, BuNo 13572 introduced the ventral rack, and BuNo 13992 was the first Goodyear's Corsair with R-2800-8W engine. Altogether, Goodyear built 2,010 aircraft of FG-1 model (FG-1A designation was never made official by BuAer), of which 410 found their way to the British Fleet Air Arm, where they served as Corsair Mk IVs (identification numbers KD161–KD570).

Beginning with BuNo 67055, Goodyear produced FG-1D, equivalent to F4U-1D. Again, it's not clear when the 'blown' frameless canopy first appeared. BuNo 76149 introduced the propeller of diameter reduced to 13 ft 1 in (3.99 m), BuNo 87788 introduced underwing rocket launchers, whilst BuNo 87988 was the first with the white recognition light removed from starboard wing upper surface. Late-production FG-1Ds featured a small fairing over the tailwheel well doors, which Vought Corsairs didn't have.

Of 1,997 FG-1Ds built, 447 were delivered to the British as Corsair Mk IVs (KD571–KD999 and KE100–KE117). Another 80 were refused by the British because of the war's end, and the last 40 (KE390–KE429) were not produced.

One FG-1 (BuNo 13041)[16] served as a flying engine test-bed for Westinghouse 19A Yankee jet engine. A small nacelle with the jet engine was mounted under fuselage. The first air test was carried out on 21st January 1944. Eight FG-1s were rebuilt as prototypes of XF2G-1 powered by R-4360 Wasp Major engine. A few other aircraft were used as test-beds during the development program of F2G Super Corsair fighter. In two of them (BuNo 14091 and 14092) fuselage spine was lowered and teardrop-shaped canopies, taken from the P-47, were installed. After the war a substantial number of FG-1Ds served, under NFG-1D designation, as trainers in US Navy reserve units.

F4U-2

The idea to create a specialized carrier-borne night fighter equipped with onboard radar was conceived in BuAer as early as August 1941. The following month the Bureau inquired Vought about constructing a night fighter version of F4U-1. In response, on 6th January 1942 Vought submitted a project of such aircraft (factory designation VS-325), and on 28th January a mock-up was accepted by a committee of US Navy specialists. Shortly afterwards the company received a formal order for a night fighter version, designated F4U-2. Since at that time Vought was busy starting the serial production of the standard F4U-1 model, technical documentation for the F4U-2 was handed over to Naval Aircraft Factory in Philadelphia. In the summer of the same year the first production F4U-1 (BuNo 02153) was delivered there and rebuilt as XF4U-2, the night fighter version prototype.

The most important modification was mounting AIA (Airborne Intercept Model A) radar, constructed at Radiation Laboratory of the famous Massachusetts Institute of Technology (MIT) and later produced by Sperry Gyroscope

F4U-1C model being tested in NACA wind tunnel at Langley in April 1943. The idea to produce a cannon-armed Corsair had been conceived quite early (which explains why the model has the old-style 'birdcage' canopy), but it wasn't put into effect until mid-1944.

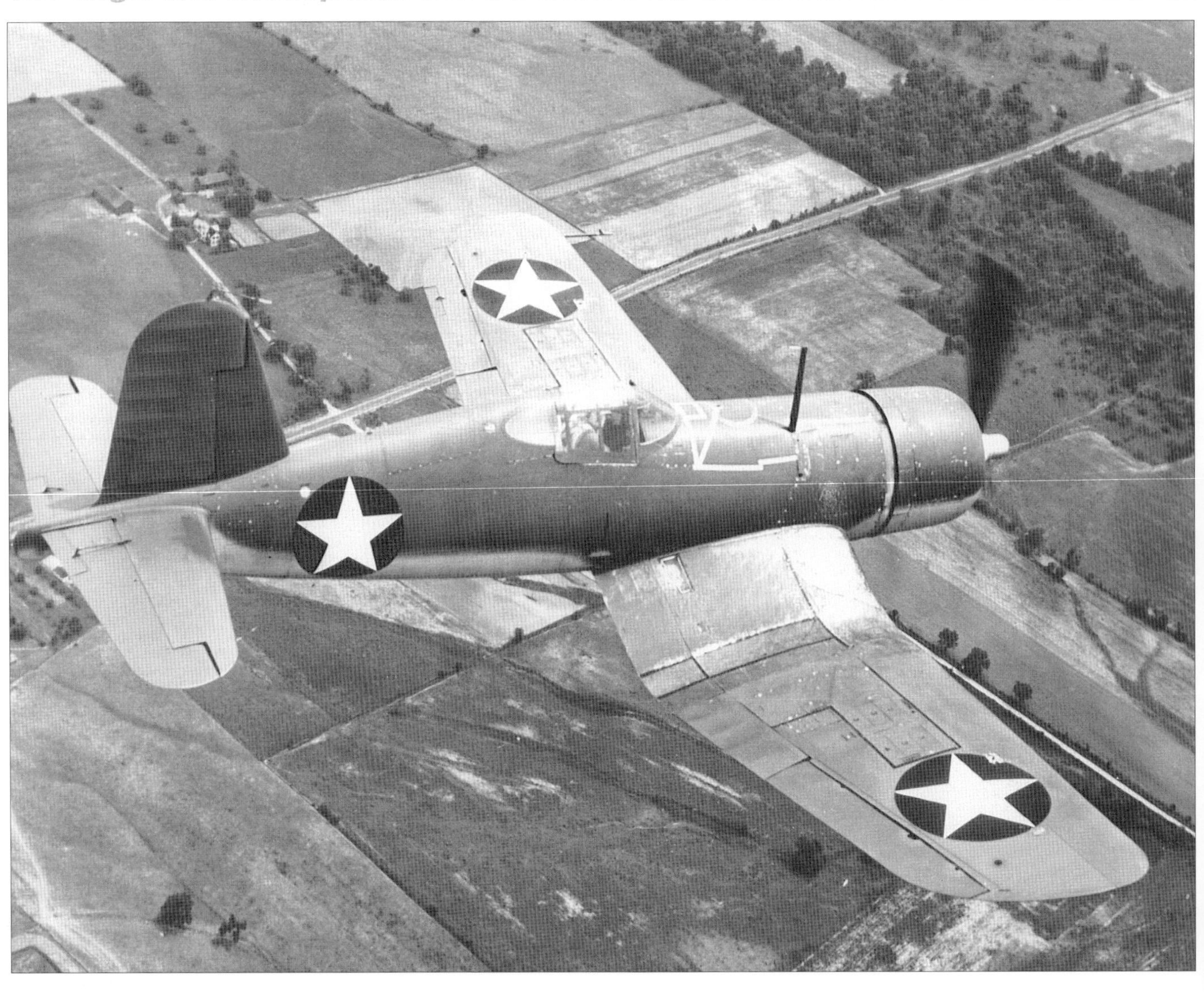

Early-production F3A-1 on a test flight. The aircraft carries the two-tone Blue Gray/Light Gray camouflage scheme used in 1942. In the period between June 1943 and July 1944 Brewster plant at Johnsville delivered to the Navy only 735 F3A-1s.

Company. The device, weighting 250 lbs (113 kg) and fitted with a parabolic antenna of 18 inches (45.7 cm) in diameter, was enclosed in a nacelle faired into the leading edge of the starboard wing, outboard of fuel tank.[17] The outboard machine gun in the starboard wing was deleted in order to counter the weight of the radome. A round 3-inch (76 mm) radar scope was mounted in the middle of the instrument panel, which necessitated relocating some of the instruments. Radar control panel was mounted on the right side of the cockpit. Cockpit lightning was modified for night operations. A more efficient generator was installed to power additional electrical equipment, and to provide cooling air for it, a small scoop was added on the right side of the forward fuselage. A radio altimeter system (AN/APN-1) was also installed, with two inverted T-shaped antennas mounted under fuselage. Standard HF radio was replaced by a new VHF set, which enabled removing the antenna mast. Furthermore, exhaust flame dampeners were fitted.

On 7th January 1943 thus equipped XF4U-2 flew to NAS Quonset Point, Rhode Island, for tests in flight. Mounting the radar resulted in slight decrease of maximum speed but handling characteristics were not appreciably affected. Thereby, 31 more F4U-1s were converted accordingly at NAF. All of them retained 'birdcage' canopy, but at least some were fitted with a taller tailwheel strut, as in F4U-1A, which came later. Arresting hooks were usually removed when the aircraft operated from land bases. The standard practice was reducing amount of ammunition to 200–250 rounds per gun to save weight.

The F4U-2 equipped two US Navy night fighter squadrons – VF(N)-75 and VF(N)-101 – and one Marine squadron – VMF(N)-532. The latter squadron converted two F4U-1As to the F4U-2 configuration (but with bubble canopies). Hence, a total of 34 F4U-2s were built. The night-fighting Corsairs served with success until January 1945. They held the distinction of being the first radar-equipped night fighters of the US Navy and the first Corsairs operating from carrier decks under combat conditions. At the end of the war F4U-2s phased out in favor of Grumman F6F Hellcat night fighter versions. After the war Vought, benefiting from experience with F4U-2, constructed F4U-4N and F4U-5N night fighters.

XF4U-3/FG-3

On 14th June 1941 BuAer turned to Vought with a proposal of constructing a high-altitude version of the Corsair. The aircraft was to be powered by an engine fitted with a two-stage turbo supercharger developed by Rudolph Birmann of Turbo Engineering Company from Trenton, New Jersey. It was expected that the turbo supercharger would allow the aircraft to maintain high maximum speed up to 40,000 ft (12,192 m). Project VS-331, developed in response to this request, was accepted by BuAer, and in March 1942 Vought was authorized to convert two F4U-1 airframes to prototypes of the new version, designated XF4U-3. On 26th December a third prototype was ordered.

Due to increasing wartime demands, Vought focused on developing and stepping up production of the standard F4U-1 model. It was not until summer 1943 that XF4U-3 came back to attention. By that time it was decided to use XR-2800-16 (C-series) engine; coupled to Birmann turbo supercharger, the engine was to maintain maximum power of 2,000 hp up to 30,000 ft (9,144 m). The turbo supercharger of 1009A type was mounted in a fairly big, streamlined fairing under fuselage, with air scoop located below engine cowling. A new four-bladed Hamilton Standard propeller of 13 ft 2 in (4.01 m) in diameter (with blades of 6501A-0 type) was installed; the same propeller was later mounted in F4U-4.

The first prototype of the new version, designated XF4U-3A, was a converted F4U-1A BuNo 17516. The aircraft featured the older, 'birdcage' canopy. It was first flown by test pilot Bill Horan on 26th March (some sources claim it was on 16th or 22nd April) 1944. The second prototype – XF4U-3B – was a converted F4U-1A BuNo 49664 airframe, with 'blown' canopy and other modifications typical of late-production F4U-1As. Due to delays in delivery of XR-2800-16 engine, a serial-production R-2800-14W with water/methanol emergency boost was used instead. Fitted with a turbo supercharger, the engine produced 2,100 hp on takeoff at 2,800 rpm and could maintain the war emergency power of 2,600 hp from sea level up to 28,200 ft (8,595 m). XF4U-3B was first flown on 20th September 1944.

The third prototype – XF4U-3C – was to be converted from the fifth production F4U-1 (BuNo 02157), but the aircraft was lost in a crash and didn't participate in the development program. Both completed XF4U-3 prototypes were delivered to the Navy in 1945 for more tests. Eventually however, because of imminent end of hostilities, the increasing production of the new F4U-4 model and ongoing technical problems with turbo superchargers, a decision was made not to develop XF4U-3 any further.

In late 1944 also Goodyear received an order for conversion of 26 FG-1D airframes to the high-altitude FG-3 model, based on XF4U-3B. This number of aircraft was to suffice for

Late-production F3A-1 in tricolor finish, with 'blown' canopy introduced by F4U-1A. Modifications typical of F4U-1D model were not incorporated into Brewster-produced Corsairs. Unlike Goodyear's FG-1, the F3A-1 had a terrible reputation in the Navy and was used mainly for training.

Three FG-1As in flight. Of note are the differences between Vought's, Brewster's and Goodyear's camouflage applying style. Altogether, Goodyear's plant at Akron, Ohio, produced 4,007 Corsairs between April 1943 and September 1945.

an entire operational squadron tasked with conducting service evaluation. The aircraft were powered by R-2800-14W engines with Birmann turbo superchargers, but otherwise they were typical FG-1Ds (equivalent to F4U-1D), which made them different from XF4U-3 prototypes. The FG-3 turned out to be the fastest of all Corsair models – 469 mph (755 kph) at 33,600 ft (10,241 m) in clean condition. First FG-3s were delivered to the Navy in July 1945.

It's not certain whether Goodyear had completed the conversion of all 26 aircraft before the program of high-altitude Corsair was cancelled. According to some sources, 13 airframes were converted, while others claim that there were no more than a few. As late as June 1946 "Naval Aviation News" wrote about plans to convert 27 (not 26) aircraft, so as to equip an entire squadron. In July 1947 two FG-3s were delivered to NATC (Naval Air Test Center) at NAS Patuxent River, Maryland, for electronic equipment tests. The last FG-3 was struck off the Navy inventory on 31st July 1949.

Other versions

During the war a number of Corsairs were modified in the field to perform tactical reconnaissance missions, by installing K-21 photographic camera in fuselage midsection. The camera was operated from the cockpit. The 'recce' Corsairs, unofficially designated F4U-1P (P for 'Photographic') participated in regular combat missions with other aircraft, documenting results.

In 1945 Goodyear proposed a night fighter version of FG-1D, designated FG-1E and based on Vought's F4U-4E project being developed at that time. The aircraft was to be fitted with AN/APS-4 radar and armed with four 20 mm cannons. However, the Navy showed no interest in the project, having sufficient numbers of Hellcat night fighters.

By all accounts, Corsair was a handful to fly, especially for inexperienced pilots. War prioritized uninterrupted production of combat aircraft and there was no time to design a training version. At the end of the war, when F4U-4 model was in production, Vought came up with an idea of converting war-weary F4U-1Ds into advanced two-seat High Speed Corsair Trainers. The second cockpit was installed in place of removed fuselage fuel tank. Two smaller fuel tanks were fitted in under cockpits, and two more in outer wing panels, in place of removed outboard and center pair of guns. Total fuel capacity decreased to 175 US gal (662 l). The provision to mount underwing pylons was retained. In 1946 Vought made such conversion of one F4U-1D (factory designation V-354). Since the Navy was uninterested, the project was dropped.

To be continued

Selected bibliography:

Bowman M.W., *Vought F4U Corsair*, Crowood Press, Ramsbury 2002.

Chance Vought F4U Corsair, „Aviation Classics" Issue 12, 2011.

Dial J.F., *The Chance Vought F4U-1 Corsair* (Profile Aircraft no. 47), Profile Publications, Leatherhead 1965.

Grossnick R.A., *United States Naval Aviation 1910–1995*, Naval Historical Center, Washington 1997.

Guyton B.T., *Whistling Death. The Test Pilot's Story of the F4U Corsair*, Orion Books, New York 1990.

Jarski A., *F4U Corsair* (Monografie lotnicze nr 11), AJ-Press, Gdansk 1993.

Jones L.S., *U.S. Naval Fighters. Navy/Marine Corps 1922 to 1980s*, Aero Publishers, Fallbrook 1977.

Kinzey B., *F4U Corsair in detail & scale. XF4U through F2G*, Part 1 (In Detail & Scale no. 55), Squadron/Signal, Carrollton 1998.

Kinzey B., *U.S. Navy and Marine Aircraft of World War II. Fighters*, Vol. 2, Revell-Monogram, Northbrook 2004.

Maki H., Tamura T., Kuroki H., *Vought F4U Corsair* (Famous Airplanes of the World no. 88), Bunrin Do, Tokio 2001.

Matt P.R., Robertson B., *United States Navy and Marine Corps Fighters 1918–1962*, Aero Publishers, Los Angeles 1962.

Moran G.P., *The Corsair and other Aeroplanes Vought 1917–1977*, SunShine House, Terre Haute 1991.

Pilots Manual for F4U Corsair, Aviation Publications, Appleton 1989.

Sturtivant R., Burrow M., *Fleet Air Arm Aircraft 1939 to 1945*, Air-Britain, Tunbridge Wells 1995.

Sullivan J., *F4U Corsair in Action* (Aircraft in Action no. 145), Squadron/Signal, Carrollton 1994.

Swanborough G., Bowers P.M., *United States Navy Aircraft since 1911*, Putnam, London 1990.

Thetford O., *British Naval Aircraft since 1912*, Putnam, London 1991.

Tillman B., *Vought F4U Corsair* (Warbird Tech Series no. 4), Specialty Press, North Branch 1996.

U.S. Navy Carrier Fighters of World War II, Squadron/Signal, Carrollton 1987.

Endnotes

[1] With time, R-2800 turned out to be a very successful, reliable and durable engine. Over 125,000 were built, in five major series and several dozen versions, between 1939 and 1960.

[2] Those wings became Corsair's most distinctive feature, earning the aircraft the name of 'Bent-Winged Bird'.

[3] The intakes were fitted with flow splitters, which at high speed produced a whistling sound generated by the air flowing through them. It was this sound that prompted Japanese soldiers to give the Corsair the ominous name of *Hyuhiyu to iu otowodasu-shi-shin* – the 'Whistling Death'.

[4] In 1943 both companies again split up, but remained subdivisions of the UAC. Vought reverted to the former name of Chance Vought Aircraft Division of United Aircraft Corp.

[5] The heavily framed cockpit canopy was popularly referred to as 'birdcage' or 'greenhouse'.

[6] Iluminated Sight Mk 8, based on the British GM2 sight. Some early production aircraft were equipped with Mk 7 reflector gunsight.

[7] The airfield was located at Stratford but was run by authorities of nearby Bridgeport, hence the name.

[8] At that time the only fighter faster than XF4U-1 was Army's twin-engined Lockheed XP-38, which reached maximum speed of 413 mph (665 kph).

[9] National Advisory Committee for Aeronautics, created in 1915, the forerunner of NASA. The facility at Langley was known as Langley Memorial Aeronautical Laboratory (LMAL).

[10] Mk III Airborne Transponder, commonly known as IFF, was a fairly simple device: after picking up a radar signal (which meant that the aircraft was in range of a radar station), it responded by transmitting a coded signal which enabled the recognition of the aircraft as 'friendly'.

[11] According to other sources, it was the fourth serial-production machine, BuNo 02156.

[12] According to some sources, XR-4360 was also fitted to BuNo 02312.

[13] Sources vary. Some claim that the new hood was introduced in BuNo 17647, while others maintain that it was BuNo 17717. It's known for a fact that most if not all of the 25 British Corsair Mk Is from this production block (BuNo 17592–17616) had the 'birdcage' type canopies. On the other hand, it's known that for example BuNo 17629 (famous 'Big Hog' flown by LtCdr John T. Blackburn, the commander of VF-17) was fitted with the new type of hood. It's possible then that new hoods were mounted beginning with BuNo 17456, but initially not in every aircraft, so that some machines with higher serial numbers still had 'birdcage' canopies.

[14] The 5-inch FFAR rockets were 5 ft 5 in (1.65 m) long and weighted 80 lbs (36 kg). Their deficiency was low speed (485 mph; 780 kph) and short range (below 1 mile; 1.6 km). They quickly phased out in favor of longer-range (3 miles; 4.8 km) HVAR rockets, also known as 'Holy Moses' – 6 ft (1.83 m) long, weighting 140 lbs (64 kg) and flying much faster (950 mph; 1530 kph).

[15] As an ammunition-saving measure, a switch at the trigger enabled firing only two, four or all six guns simultaneously.

[16] BuNo 13044 according to other sources.

[17] The radar's wavelength was 3.2 cm. It could detect big airborne objects (like bombers) from up to 8,000 yards (7,300 m), but its practical range was about only three miles (4,800 m). It scanned through 120° cone (±60° from aircraft axis).

In early 1944 this FG-1 (BuNo 13041 or 13044?) served as a flying engine test-bed for Westinghouse 19A Yankee jet engine. A small nacelle with the jet engine was mounted under fuselage.

Camouflage and markings

US Navy, 1942–1945

The XF4U-1 prototype had a typical prewar painting scheme and markings. All surfaces (both metal- and fabric-covered) were finished in aluminum dope (FS17178), with only wing upper surfaces painted with ANA506 Orange (Chrome) Yellow (FS13538).[1] The yellow on wing leading edges slightly overlapped wing undersides. National markings, in form of a five-pointed white star superimposed on a circular dark blue background and carrying a small red disc in its center, were painted on upper and lower surfaces of both wings with ANA511 Insignia White (FS17875), ANA502 Insignia Blue (FS15044) and ANA509 Insignia Red (FS11136). 'U.S. NAVY' was stenciled in block letters on either side of rear fuselage, and the fin and rudder carried the serial number '1443' and model designation 'XF4U-1', respectively – all inscriptions in ANA515 Black (FS17038). Propeller blade tips featured three warning stripes, each 4 inches (10.2 cm) wide, in Insignia Red, Orange Yellow and Insignia Blue (from the tip down). All paints were glossy except for ANA604 Non-Specular Black (FS37038), which partially covered backsides of propeller blades.

Before F4U-1 entered production and saw operational service, camouflage and markings

This NFG-1D (BuNo 92041) served for training at NAS Livermoore, California, in early 1946. Many FG-1Ds, withdrawn from combat units after the war, served under NFG-1D designation as trainers in US Navy reserve units.

of US Navy aircraft had radically changed. Upper surfaces and sides were painted in Non-Specular Blue Gray[2] (approximate to FS35189), and undersides in ANA602 Non-Specular Light Gray (FS36440). Undersides of the outer, foldable wing panels (visible from above after folding) were also painted Blue Gray. National insignia, by then devoid of the red discs inside white stars, were painted on upper and lower surfaces of both wings, and on either side of rear fuselage, with ANA601 NS Insignia White (FS37875) and ANA605 NS Insignia Blue (FS35044).

Vertical fin carried branch of service (NAVY or MARINES) lettering and Bureau Number, while the model designation was painted on the rudder; they were all 1 inch (25.4 mm) high. Fuselage tactical markings were 12 inches (30.5 cm) high and in full version they consisted of a squadron number, separated by a dash from a letter designating the type (class) of squadron[3], separated by a dash from the aircraft's individual number (for example, 17-F-5 read as aircraft no. 5 of VF-17). Sometimes the fuselage markings were limited to squadron type letter and the individual number (not separated by a dash), or even to the aircraft's number only, which was often repeated on main landing gear wheel covers and/or on either side of the engine cowl. All numbers and letterings were painted ANA604 NS Black or ANA601 NS Insignia White. Both sides of propeller blades were painted ANA604 NS Black. The tricolor warning stripes on propeller blades continued to be painted in glossy ANA509 Insignia Red, ANA506 Orange Yellow and ANA502 Insignia Blue.

XF4U-3A prototype, originally F4U-1A BuNo 17516 converted to the high-altitude version fitted with XR-2800-16 engine and Birmann turbo supercharger. Of note is the old-style 'birdcage' canopy. The turbo supercharger intake was located under engine cowl.

XF4U-3B, the second prototype of the high-altitude version, was a converted F4U-1A BuNo 49664 with 'blown' canopy and other modifications typical of late-production F4U-1As. The aircraft was powered by R-2800-14W engine with turbo supercharger.

XF4U-3B (BuNo 49664) with folded wings. Due to imminent end of hostilities, the increasing production of the new F4U-4 model and ongoing technical problems with turbo superchargers, a decision was made not to develop XF4U-3 any further.

The next change in camouflage and markings took place on 1st February 1943, after Specification SR-2c had been issued by BuAer on 5th January. The new basic non-specular tricolor camouflage scheme outlined in SR-2c was as follows: upper, horizontal surfaces of wings and tail planes were to be finished in ANA606 Semi-Gloss Sea Blue (FS25042); top fuselage and leading edge upper surfaces of wings and tail planes, extending back approx. 5% of the chord – in ANA607 NS Sea Blue (FS35042); undersides of fuselage, wings and tail planes – in ANA601 NS Insignia White; vertical fin, rudder and undersides of outer wing panels (exposed to view from above when folded up) – in ANA608 NS Intermediate Blue (FS35164). Fuselage sides were to be painted in a manner assuring smooth transition between the Sea Blue on the upper surfaces and White on the undersides. In practice, fuselage sides were simply painted in ANA608 Intermediate Blue. In case of Corsairs, fuselage sides above wings were finished in ANA607 NS Sea Blue.

The same specification introduced changes in location of the national insignia. Now they were to be applied only to port wing upper surface, starboard wing underside and both fuselage sides. Stenciling on fin and rudder and tactical markings on the fuselage were to be painted Black on Intermediate Blue background, but in practice White tactical markings were common. Besides, tactical markings were increasingly often reduced to aircraft's individual number, repeated on either side of the engine cowl and (in Black) on main landing gear covers. Propeller blades remained ANA604 NS Black, with band of ANA614 NS Orange Yellow (FS33538), 4 inches (10.2 cm) from tip toward the hub.

Army-Navy Specification AN-I-9a, issued on 29th June 1943, modified national insignia. White horizontal rectangles (bars) were added to the sides of the blue circle with star. The en-

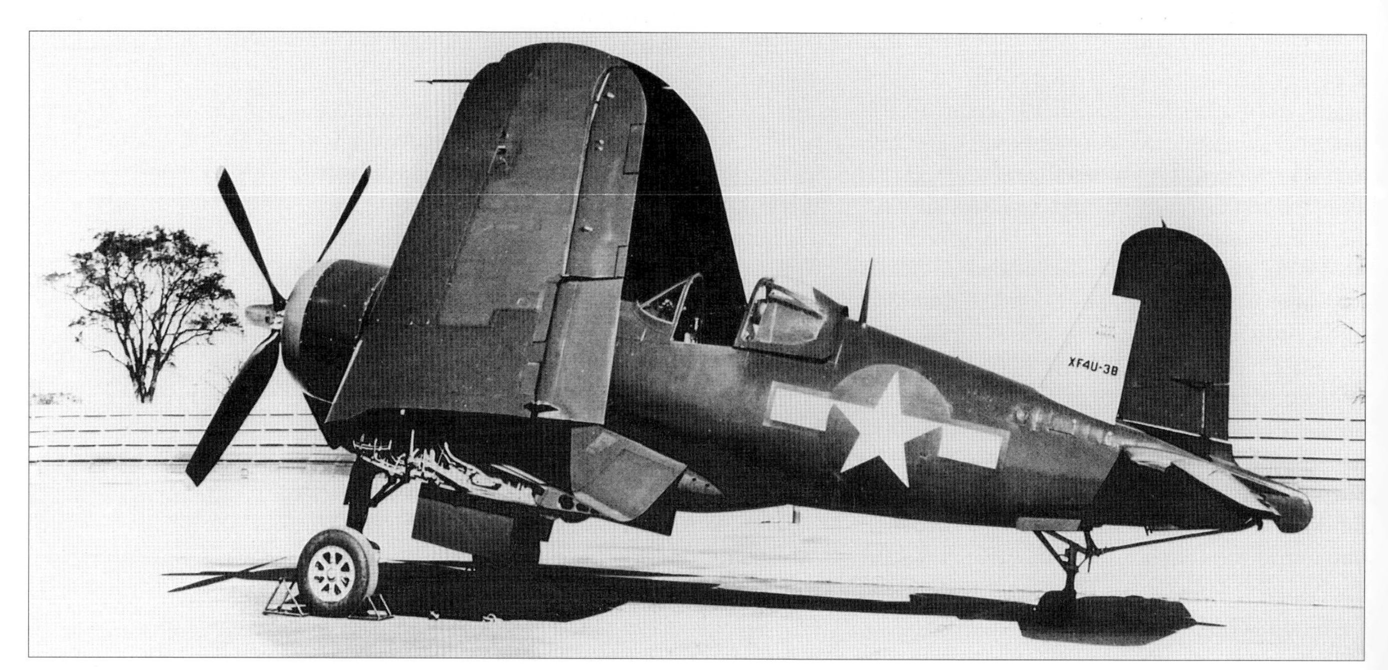

tire insignia was then outlined with a border of Insignia Red.[4] Only a few weeks later, on 14th August, updated Specification AN-I-9b ordered changing the color of the outline to Insignia Blue. In this form the US national insignia remained unchanged until early 1947.

On 22nd December 1943 BuAer issued Specification SR-2d, with an effective date of 6th March 1944. Tactical markings were reduced to aircraft's individual number. These numerals were to be painted in white on both sides of the vertical stabilizer, above tail planes, and measure 16 inches (40.6 cm). The numerals were repeated on the sides of engine cowl and undercarriage covers of 6 inches (15.2 cm) high. However, some units continued to use fuselage markings in a full form, as before.

The last change in Corsairs' camouflage came about with the issuing of Amendment 1 to Specification SR-2d, on 13th March 1944. The amendment specified that all fighter aircraft be camouflaged in ANA623 Glossy Sea Blue (FS15042) over all exterior surfaces. National insignia were also painted in glossy colors – ANA511 Insignia White (FS17875) and ANA502 Insignia Blue (FS15044). All numerals and letterings were painted ANA511 Glossy Insignia White.

Specification SR-2e, issued on 26th June 1944, authorized applying an antiglare panel on top of the fuselage ahead of the windshield, which was to be painted in ANA607 NS Sea Blue. Beginning with 10th March 1945, when Amendment 2 to SR-2e was issued, propeller hubs were to be finished in ANA604 NS Black.

On 27th January 1945 BuAer issued a document known as Air Force, Pacific Fleet, Confidential Technical Letter No. 2 CTL-45. It standardized the system of geometric recognition symbols on aircraft operating from fleet and light carriers (CVs and CVLs) of the Pacific Fleet. The G-symbols identified which carrier (and Carrier Air Group, CVG) each aircraft belonged to. The geometric figures or stripes were applied to wingtips or ailerons (on port wing underside and on starboard wing upper surface), and on both sides of vertical fin and/or rudder. Glossy paints were used, usually ANA511 Insignia White, sometimes ANA506 Orange Yellow or ANA505 Light Yellow (FS13655), and only rarely ANA503 Light Green (FS14187). On 2nd June Air Force, Pacific Fleet, Confidential Technical Letter No. 4 CTL-45 introduced similar recognition G-symbols on aircraft stationed at escort carriers (CVEs) of the Pacific Fleet.

Another change in the recognition system of aircraft stationed at CV and CVL carriers took place shortly before the end of the war, on 27th July 1945. The commander of Task Force 38 (TF-38) issued a dispatch No. 061121, in which he pointed out that the geometric symbols then in use were not intelligible visually and they were hard to describe over the radio. He recommended using a single- or double-letter codes instead. The codes were applied with ANA511 Glossy Insignia White paint in the same places as the G-symbols: on port wing underside, starboard wing upper surface, and on both sides of fin and rudder. The block capital letters were 24 inches (61 cm) in height. A month later to the day, in dispatch No. C2CTF/F39 (01614), the commander of TF-38 ordered relocating aircraft's individual numbers back to the fuselage, so that only letter codes could remain on the vertical tail surfaces.

One more shot of XF4U-3B (BuNo 49664) prototype. The aircraft was first flown on 20th September 1944. Of note is the new four-bladed Hamilton Standard propeller of 13 ft 2 in (4.01 m) in diameter, also used in F4U-4 model.

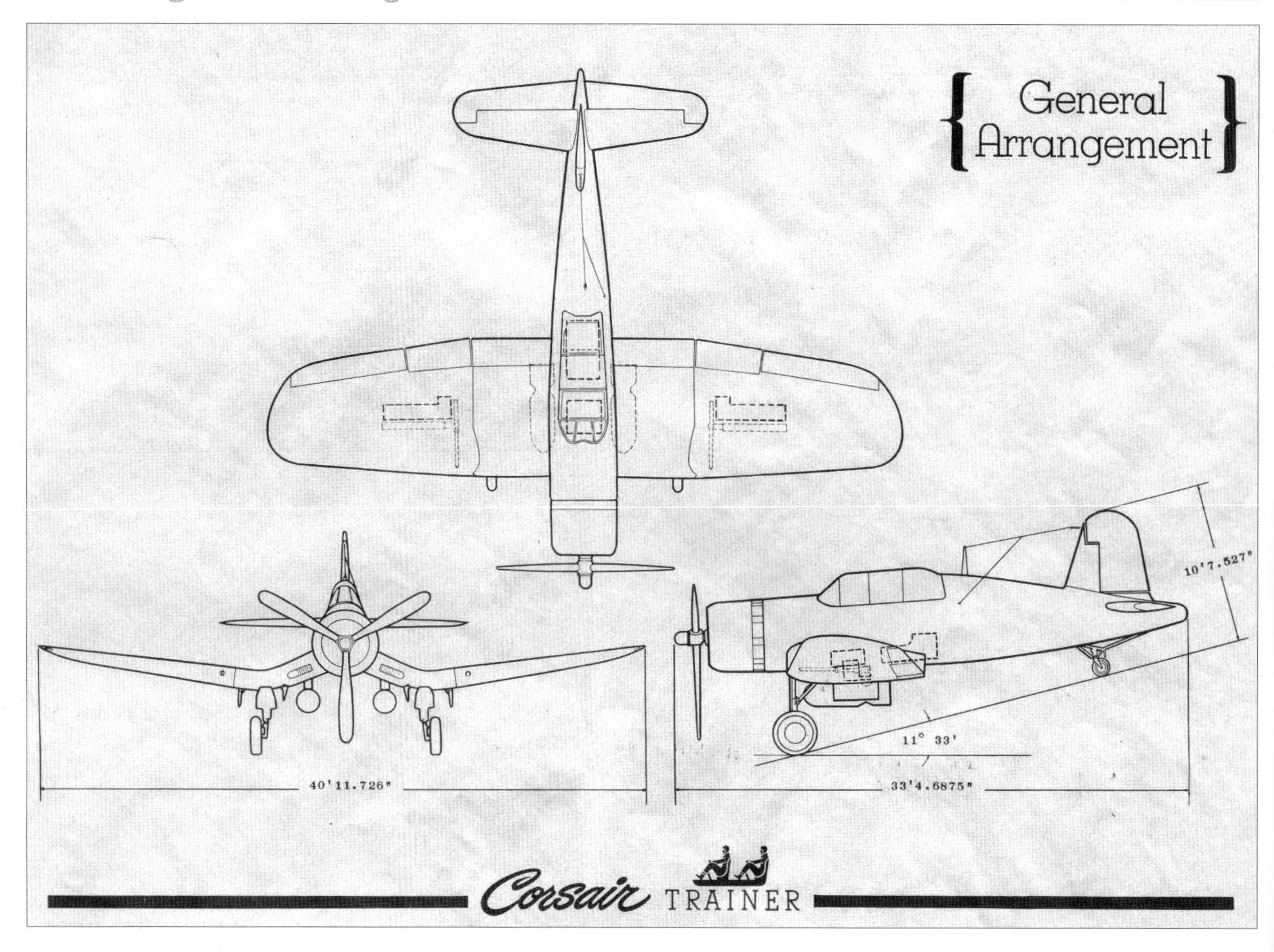

Vought's tender drawing demonstrating the silhouette and basic dimensions of the proposed two-seat High Speed Corsair Trainer. Since the Navy was uninterested in turning war-weary F4U-1Ds into trainers, the project was abandoned.

On 11th February 1945 a separate system of identification was introduced for aircraft from Hawaii-based Carrier Air Service Units (CASU). As stated in A3/FF12-5/(112tr, 3133) directive, the aircraft were to carry a letter identifying their air station, followed by a number from 1 to 99 inclusive. These ID codes were applied with ANA511 Glossy Insignia White on both sides of the fuselage and on port wing underside. The codes were 36 inches (91.5 cm) high. On 10th September another directive, AF/FF12-5/(112-Cs, 19030), effective from 1st October 1945, changed the assignment of letters to units and air stations, and authorized the use of numbers over 100.

Selected bibliography:

Doll T.E., *US Navy Aircraft Camouflage & Markings 1940–1945*, Squadron/Signal, Carrollton 2003.

Doll T.E., Jackson B.R., Riley W.A., *Navy Air Colors. United States Navy, Marine Corps, and Coast Guard Aircraft Camouflage & Markings 1911–1945*, Vol. 1, Squadron/Signal, Carrollton 1983.

Elliott J.M., *The Official Monogram US Navy & Marine Corps Aircraft Color Guide, 1940–1949*, Part 2, Monogram Aviation Publications, Boylston 1989.

Sullivan J., *F4U Corsair in Color*, Squadron/Signal, Carrollton 1981.

Endnotes

[1] ANA numbers for Non-Specular and Semi-Gloss paints were introduced on 28th September 1943, when Army-Navy Aeronautical Bulletin No. 157 was published, and for Glossy paints on 4th December 1943 (ANA Bulletin No. 166). However, since they are commonly referred to in other publications and served to identify many paints that had been in use before the two bulletins were published, they are cited here as well. The numbers from a modern Federal Standard 595 (FS) color chip set were given due to their popularity among modelers.

[2] Blue Gray paint was not included in any of the ANA bulletins, since at that time it was no longer in use.

[3] For example, F letter denoted all US Navy fighter squadrons (VF), MF letters – US Marine Corps fighter squadrons (VMF), BF – US Navy fighter-bomber squadrons (VBF), OF – US Navy observation-fighter squadrons (VOF).

[4] The two Insignia White bars were placed in such a position that the top of the bars formed a straight line with the top of the star's horizontal arms. The dimensions of the two bars were to be the same length as the radius of the circle with the star, and half the radius in width. The red outline was one eighth the width of the radius.

Frontline service

Campaign in the Solomons (February - December 1943)

Before F4U Corsair entered frontline service, the initiative in the Pacific war had shifted to the Allies. In the latter part of 1942, after the victorious battles of the Coral Sea and Midway, the Americans rode their momentum, landing at Guadalcanal in August 1942. A furious battle followed, which went on unabated until early February 1943, when the last Japanese troops pulled out. Guadalcanal became a bridgehead at the southeastern edge of the Solomons – an archipelago stretching out to the northwest for over 600 miles, which in the ensuing months formed the axis of the Allied advance in the South Pacific. Capturing islands was the job of the USMC (United States Marine Corps), and it was with the Marines, who needed a successor for their aging F4F Wildcats, that Corsair first saw action.

The difference between the docile-looking Wildcat and the beastly, unforgiving Corsair was enormous, and while the former had some advantages over the latter, most Marine pilots quickly learned to appreciate the potential of their new mount. Col. Jefferson DeBlanc, Medal of Honor winner, who flew both types in combat (with VMF-112 and -422), reflected:

"The transition from the Wildcat to the Corsair was rewarding. The Wildcat was a good fighter. I found it to be better platform for gunnery. It also had immediate and total forward visibility since it was primarily a carrier plane. (...) The Corsair, on the other hand, had a wide landing gear, which made for a more stable landing. But the price to be paid was a loss of forward visibility on take-off. This was due to the high angle of the fuselage. You had to properly align the flight path of the craft with the runway or you would end up in the trees.

(...) One odd feature of this Corsair was its lack of a cockpit floor. The pilot placed his feet on runners, but below was a dark, empty well. If we flew upside down and held that position, trash from below would fall on us! We used to joke about being hit by a tool left by some mechanic. Later models had a floor".[1]

Lt.Col. Harold Langstaff, who served with VMF-215 *Fighting Corsairs* and completed three combat tours in the Solomons, commented:

„The early models had a number of mechanical problems such as gas leaks in the wing tanks, and the two-thousand-horsepower engine caused a lot of vibration, which caused the exhaust stacks to crack. With fourteen-foot propellers, there was a great deal of torque created during takeoff. The pilot had to crank in the full right rudder tab and stand on the right rudder pedal to keep the plane straight. Otherwise it would pull you right off the runway to the left".[2]

F4U-1 Corsair of VMF-215 in flight near Hawaii, prior to deployment to South Pacific, January 1943.

F4U Corsairs of VMF-124 standing on the flight line on Guadalcanal, April 1943.

The first Marine fighter unit to receive Corsairs (in October 1942) was VMF-124 under Maj. William Gise. On **14th February**, two days after arriving at Guadalcanal, the unit had a first taste of combat. On that day the AirSols mounted a raid against Kahili, a major Japanese airbase on Bougainville, on the other side of the Solomons. Nine B-24s were escorted by a mixed bag of RNZAF Kittyhawks, the Army's P-38 Lightnings, and a dozen of VMF-124's Corsairs, stacked up for optimum performance. Intervening A6M Zero fighters of the IJNAF (Imperial Japanese Navy Air Force), also known as Zekes to the Americans, shot down two bombers, four P-38s, two Kittyhawks and two F4Us. VMF-124 lost two pilots killed in action. Lt. Lyon collided head-on with a Japanese fighter. Lt. Stewart ran out of fuel due to a punctured tank and had to ditch at sea; the Zero pilots shot him dead in the water.

On **13th March**, when the Allies mounted a raid against Japanese airfields at Vila (on Kolombangara Island) and Munda (on New Georgia), losses were just as heavy. Perhaps fortunately for the embryonic Corsair force in the Solomons, for the reminder of March and most of April 1943 it was only rarely required to tangle with the Japanese in the air, otherwise it could have been wiped out. Instead, strafing missions were frequent. For example, on **28th** and **29th March**, VMF-124 attacked Japanese seaplane bases at Shortland Islands (Poporang and Faisi) in the upper Solomons.

Meanwhile, in late February the Marines captured Russell Islands located some 50 miles northwest of Guadalcanal. Shortly afterwards they began to construct an airstrip at Banika, one of the Russells. On **1st April** the Japanese set out in force to investigate. The raiding party comprised D3A Val dive bombers and 58 Zeros, drawn from the 204th, 251st and 582nd Kokutais (Air Groups). The usual mixed formation scrambled to contest this incursion, 42 fighters in all, most of them F4F Wildcats, but also eight VMF-124 Corsairs and six P-38s. 2/Lt. Kenneth Walsh of VMF-124 scored his three premier victories: two Zeros and a Val. The air battle went on for nearly three hours; nonetheless, the losses on both sides were surprisingly low, a dozen or so aircraft in all (the Americans lost five Wildcats and one Corsair).

On **3rd April** VMF-213 *Hellhawks* arrived at Guadalcanal, relieving VMF-124, which had just completed its first combat tour in the Solomons and left for R&R (Rest and Recuperation) in Sidney (combat tours were about six weeks long; squadrons returning from R&R usually spend a few more weeks in the rear area, at Espiritu Santo Island, gearing up for combat).

During the first few days after their arrival at Guadalcanal, *Hellhawks* performed CAPs (combat air patrols) over Banika and twice escorted photo-recce B-24s to New Georgia. Somehow they missed a massive air battle fought in the Russells area on **7th April**. Only Lt. Herman Spoede succeeded in shooting down a lone Zero during a routine CAP, winning the first victory for his squadron.

On **13th April** Maj. Wade Britt, the CO of VMF-213, was killed in a fatal crash. Taking off in a hurry to catch up with the rest of his squadron, he ran off the runway and crashed into two Corsairs that were parked on the taxiway, as a

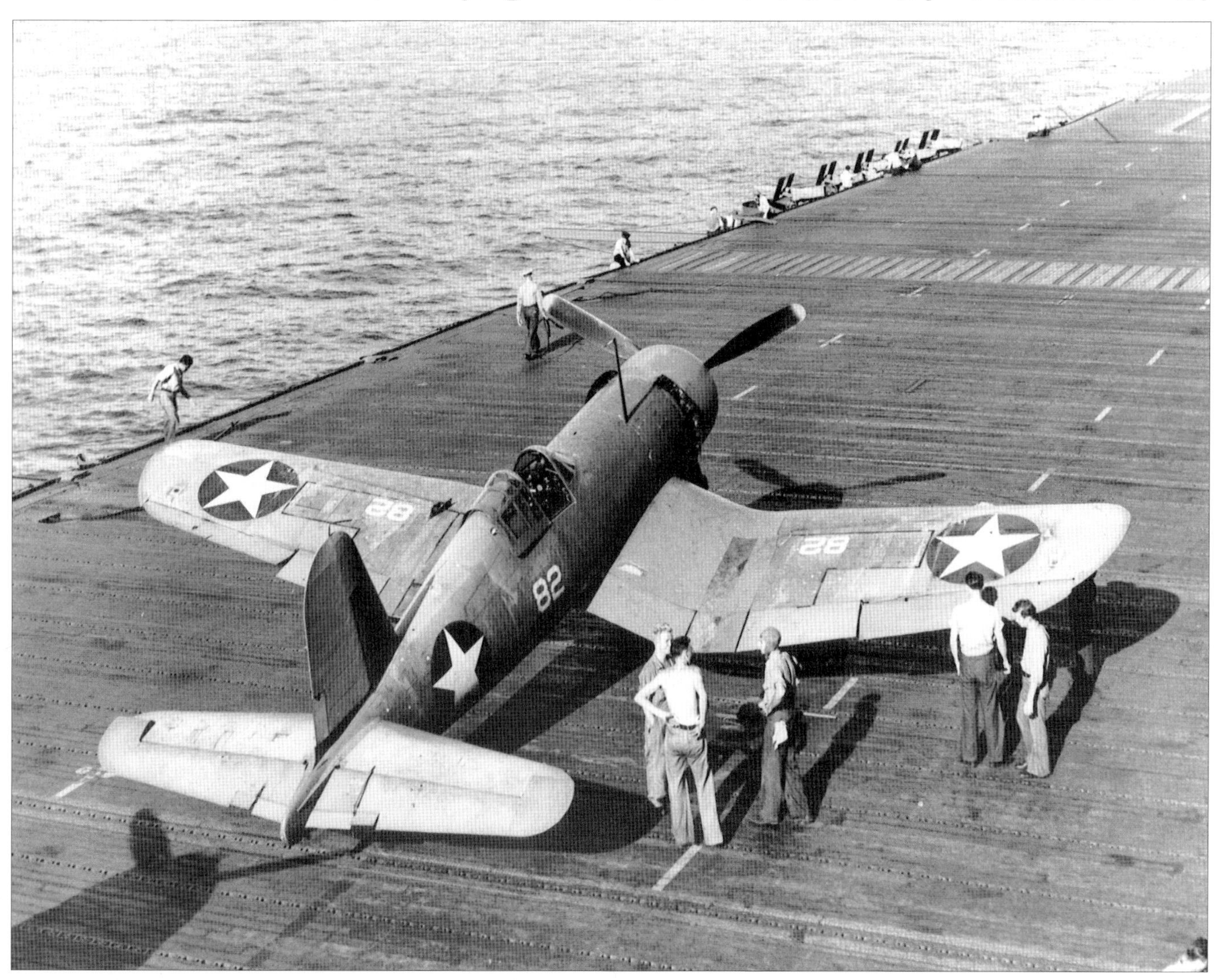

This F4U-1 was briefly assigned to USS Enterprise (CV-6) for carrier evaluation in early 1943. It was generally flown by Lt. Stanley "Swede" Vejtasa, the Ops Officer of VF-10, the squadron stationed aboard Enterprise.

F4U-1 Corsairs of VF-12 in flight, March 1943.

result of which all three planes burst into flames and were completely destroyed. The squadron was taken over by Maj. Gregory Weissenberger. The following days VMF-213 flew routine missions, often escorting photo-recon aircraft to New Georgia and Rekata Bay at the coast of Santa Isabel Island, where the Japanese had a seaplane base. On **22nd April** the squadron carried out a hazardous strafing mission against Japanese airfields at Munda (on New Georgia) and Vila (on Kolombangara).

It was not until **25th April** that VMF-213 again engaged enemy in the air. In the morning 12 Corsairs took off from Guadalcanal to escort

One of VF-17's F4U-1s on the USS Charger during carrier landing qualifications in Chesapeake Bay, March 1943.

six Marine SBD Dauntless dive bombers on a bombing mission against the airfield at Vila. Four of the Corsairs were to make a strafing attack on the village of Vanga Vanga on Kolombangara. Over Vila the strafing flight broke off from the main formation and went down to work over their assigned target. As the four pilots started homeward, south of Vangunu Island they spotted the enemy. "There were approximately 12-16 twin engined monoplanes Jap bombers, which appeared to be black. Above the bombers there were three layers of cover furnished by Zeros"[3].

Climbing under cover of clouds, the four Corsairs were first sighted when on level with the bombers and were immediately attacked by 16-20 Zeros (the Zeros, before making their attack, dropped their belly tanks in unison). The fight lasted only a few minutes. The four *Hellhawks*, despite overwhelming odds, succeeded in breaking up the raid and turning the enemy back. Moreover, they shot down six Zeros (Maj. Peyton three, Lt. Vedder two, Lt. Peck one) as against the loss of two aircraft and one pilot. Maj. Monfurd Peyton returned with wound in his left arm and left knee, and in a badly shot-up plane. Lt. Milton Vedder was shot down, bailed out and landed safely in the water; two days later he was returned to his unit.

Around that time the motley array of allied air units at Guadalcanal, hitherto known as the 'Cactus Air Force', was superseded and absorbed by the Air Solomons. The AirSols, as it was officially abbreviated, put all air forces operating in the area under one command, and therefore encompassed the Navy, the United States Marine Corps (USMC), the United States Army Air Force (USAAF) and the Royal New Zealand Air Force (RNZAF). More Corsair squadrons were soon to enter the fray in the Solomons, among them VMF-112 *Wolfpack* and VMF-221 *Fighting Falcons*, which converted to the type in May.

In May the air war over the Solomons almost fizzled out. VMF-213 *Hellhawks* continued to fly escort missions and strafed the enemy on land and at sea, but failed to encounter opposition in the air right up to the end of its first tour. The following note in the squadron's diary, dated to **10th May**, speaks for itself: „A big formation was sighted, and for a moment, the F4U's were all primed for the fun, but soon realized that it was our bombers with their escort". The following day *Hellhawks* were relieved by VMF-124 returning to frontline duties.

Meanwhile the Navy, also in need of a successor for the F4F Wildcat, was increasingly disappointed with the Corsair. The first Navy fighter unit converted to the type, in October 1942, was VF-12. Next in line was VF-17, which received the new fighters in February 1943. Early-production Corsairs, quite reliable when operated from land bases, proved too troublesome for carrier operations (most of the problems were related to making arrested landings). In mid-1943 the Navy lost patience and declared Corsair unfit for its designed role of a carrier-borne fighter – at least for the time being. VF-12, which had lost seven pilots killed during training accidents on Corsairs, re-armed with F6F Hellcats in July 1943. Most notably, LtCdr John 'Tom' Blackburn, the commanding officer of VF-17 *Jolly Rogers*, resolved to carry on with Corsairs. In a curious sequence of events, the Navy conceded, but when the squadron finally mastered landings on carrier decks (by that time

VF-17 *Jolly Rogers* struggling to tame the 'Hog', as the Corsair came to be known in the Navy; USS Bunker Hill, July 1943.

VMF-213 picked up their first Corsairs off New Caledonia straight from the deck of escort carrier USS Copahee, which brought the planes from the USA to South Pacific; this F4U-1 is about to be catapulted into the air, 29th March 1943.

it operated the improved F4U-1A model) and sailed out to war aboard USS Bunker Hill, it was ordered to disembark enroute. As the only carrier-based squadron of the Navy equipped with the Corsair, VF-17 became a logistical headache. By that time Marines operated plenty of Corsairs in the Solomons, and spare parts and replacement aircraft were being directed there. So off to the Solomons went VF-17. However, that was still months away.

On **13th May** VMF-112 and -124 intercepted a Japanese foray towards Guadalcanal. VMF-112 diary related: "The enemy was a Type 01 bomber, out on reconnaissance, escorted by 25 Nagoya Type Zeros. The reconnaissance plane never got in closer than the western tip of the Florida Islands. Capt. Donahue's flight and Lt. Maas's flight made contact as the enemy was fighting with the P-40 Knucklehead [Banika] patrol ten miles North West of the Russells".

In all, 16 Zeros were shot down: seven by VMF-112, eight by VMF-124, and one by a P-38 pilot. Own losses amounted to three Corsairs shot down (another made a forced landing). Maj. William Gise, the CO of VMF-124, was shot down and killed, possibly by *Joto Hiko Heiso* Ryoji Ohara of 204th Kokutai, one of the IJNAF aces. Kenneth Walsh of VMF-124 shot down three Zeros, becoming the first ace of his squadron and the first Corsair ace (with five victories on the type). Also VMF-112 had the first ace among its ranks – Capt. Archie Donahue added four Zeros to his tally of two victories scored on Wildcats.

VMF-112 War Diary: "Two P-40s were mistaken for zeros by F4U and had a number of holes in them, much to the discomfort of the P-40 pilots. After the engagement the P-40 pilots were over to find out who had attacked them and to say that they were excited is hardly the word. Needless to say no one in VMF-112 or VMF-124 would admit committing such an error".

In June two more Marine squadrons switched from Wildcats to Corsairs, VMF-122 and VMF-214 (the first squadron to bear that designation, also known as the *Swashbucklers* – not to be confused with Boyington's *Black Sheep*).

As the build-up of air assets continued on either end of the Solomons, pace of operations quickened. On **5th June** VMF-112 and -124 (20 Corsairs in all), aided by 24 P-40s and 5 Lightnings, escorted 15 SBDs and 12 TBF Avenger torpedo bombers targeting enemy shipping off Bougainville. VMF-112 war diary relates: "During the retirement scattered enemy forces totaling 20 to 25 float biplanes and 15 Zero fighters attacked our forces. Seven float biplanes and five Zero fighters were shot down". Lt. Wallace Sigler and Lt. Stanley Synar, two future aces of VMF-112, each shot down two float biplanes, recognized as either Pete (Mitsubishi F1M) or Dave (Nakajima E8N). In VMF-124, 2/Lt. Ken Walsh was credited with a double victory (Zero and a float biplane apiece).

Shortly before noon on **7th June** AirSols fighters intercepted a Japanese raid comprising

VMF-213 ground crew by 'Eightball' / 'Dangerous Dan', one of the squadron's Corsairs; Guadalcanal, 1943.

10 B5N Kate torpedo bombers and over 100 Zeros, heading for Guadalcanal. The defending force consisted of Warhawks (of 18th FG), Lightnings (of 347th FG), Wildcats (VF-11), Kittyhawks (No 15 Sqn RNZAF) and Corsairs of VMF-112.

Corsair pilots were credited with eight Zeros, including two by Capt. Archie Donahue, who scored his last (eighth and ninth) victories over the Solomons. On the debit side, four Corsairs were lost. All four pilots survived, although in two cases it was nothing short of a miracle. Lt. James Percy had 'made ace' when his outfit had been flying Wildcats. On that day he scored his last and sixth victory (the only one in a Corsair), before he got himself into trouble. VMF-112 war diary:

"Lieutenant Percy became separated and was attacked by at least six Zeros. He fought until his controls were shot away then he jumped at a very low altitude. Because of the low altitude his parachute streamed behind him but did not open and it is a miracle he is still alive. He hit the water sustaining a broken pelvis, two sprained ankles and numerous bruises. He was able to swim to a coral reef 5 miles east of the Russell's strip where he remained all afternoon and night, finally reaching the adjacent island where he was found by natives and returned to the Russells that afternoon."

Lt. Sam Logan's ordeal was even worse: "Lieutenant Logan had been proceeding alone, got into the second fight when he spotted a Zero on the tail of a P-40. He attacked the Zero and drove it away but in so doing he was hit from behind and the controls of his plane were shot away and he had to bail out at 18,000 feet. In this helpless altitude, while descending in the parachute, the enemy exhibited the most base cruelty by making repeated firing runs on Lieutenant Logan. On the third run Logan was trying to spill his chute and the enemy, evidently miscalculating, hit Logan with his propeller and cut half of his right foot away. The enemy still persisted in these attacks until Logan's plight was observed by Flight Leader Herrich (RNZAF) who drove the enemy away in spite of the fact that his supply of ammunition was exhausted. Lieutenant Logan showed great presence of mind in applying a tourniquet, injecting morphine and taking sulfathiazole tablets. Upon reaching the

F4U Corsairs of VMF-213 at Henderson Field, Guadalcanal, June 1943.

surface of the water midway between Buraku and the Russells, he inflated his boat, dusted his wound with sulfanilamide powder and displayed sea markers to aid his rescue. He was picked up very shortly by the J2F-5 [Duck] piloted by Lt.Col. Clifford of MAG-21. It was found necessary to amputate the right foot on his arrival at the hospital."

Another Corsair pilot who got knocked down in the course of the same engagement was Maj. Robert Fraser, the CO of VMF-112. Before he went down, Frazer had scored two victories (his fifth and sixth). He was likely shot down by *Joto Hiko Heiso* Hiroyoshi Nishizawa of 251st Kokutai, known as 'the Devil of Rabaul', one of the IJNAF top-scoring aces (on that day Nishizawa was credited with his first Corsair).

On **10th June** Japanese aircraft attacked an allied convoy off Malalita island. While defending the ships, Lt. William Crowe of VMF-124 shot down two twin-engined G4M Betty bombers (Before his combat tour was over at the end of August, Crowe's tally stood at seven victories).

Another turning point in the Solomons campaign was the air battle fought on **16th June**. The attacking force consisted of 24 Vals shepherded by 70 Zeros. Their target was allied shipping off Guadalcanal. Most victories were racked up by Navy Wildcats (31) as well as USAAF and RNZAF fighters. Lt. Howard Finn, the only pilot of VMF-124 to contact enemy, shot down a Zero and a Val apiece, near Savo Island. VMF-122 got one victory, but lost a pilot. It was the last Japanese bomber raid sent down

The pilot of this F4U-1 was variously identified as Maj. Gregory Weissenberger of VMF-213 and Maj. Joseph Reinburg of VMF-122; June 1943.

Captain Francis E. Pierce Jr of VMF-121, Guadalcanal 1943. Pierce had scored four Wildcat victories before his squadron transitioned to Corsairs. He won his fifth and last victory, a Betty bomber shot down between Rendova and Tetopari, on 30th June 1943.

to Guadalcanal. Vice Admiral Kusaka, the commander of the 11th IJNAF Air Fleet stationed at Rabaul (and later of all Japanese naval forces in New Guinea and the Solomon Islands area) found these incursions too costly.

Notably, on the same day VMF-121 left Guadalcanal for the newly completed airstrip at Banika in the Russell Islands, the first allied fighter squadron to do so. The following day VMF-213 arrived from Espiritu Santo to Guadalcanal for another combat tour, relieving VMF-124. The *Hellhawks* had to wait until the last day of the month to see some action in the air.

On **20th June** the Americans carried out their first landing on New Georgia, at Segi – practically unopposed. They immediately set about constructing a fighter airstrip. The main, bloody battle with the 10,000-strong Japanese garrison was to be fought on the other side of the island, around an airfield at Munda Point.

On **30th June** the Marines landed on the island of Rendova, opposite Munda. A heavy enemy counteraction was expected – allied aerial photoreconnaissance detected some 250 aircraft at Rabaul, and 90 more at Bougainville. Standing patrols of 32 allied fighters were maintained over the landing zone throughout the day. The Japanese lived up to expectations.

The air battle raged on from just before noon until evening. The core and majority of the defending force was, for the first time, Corsair squadrons: VMF-121, -122, -213 and -221. The top scorers were VMF-213 with 20 victories: 11 Zeros and 9 float biplanes. Lt. Wilbur Thomas shot down four Zeros – in his first combat! The same day marked the beginning of another illustrious career – Lt . Edward Shaw of the *Hellhawks* bagged three floatplanes in the vicinity of Rendova. The commander of VMF-213, Maj. Gregory Weissenberger, knocked down three Zeros, but had his Corsair shot out from under him. His account of the action is as follows:

"I reported on station to Vega [ground controller] at Rendova with six planes at about 1450 and orbited Rendova Harbor at 17,000 feet. About 1530 Vega reported a bogey approaching from the NW and sent six planes to 25,000 feet. After reaching that altitude Vega reported the enemy approaching our shipping and ordered all fighters to protect the ships.

right on going and then I saw a cloud for which I headed. As I was diving I saw oil going over the top of my hood".

Moments later Segal ditched his plane off Rendova. Upon impact with water he hit the instrument panel with his face, breaking his nose and knocking off two teeth. He spent the night floating around, until he was picked up by destroyer USS Taylor the next morning.

Segal probably didn't realize that at some point Swett shot a Zero off his tail. Swett reported:

"When I came out of the cloud I saw a Corsair which was smoking badly. I was two or three miles astern of his beam. There were three Zeros about a mile behind him in a spread-out formation. I caught up to the last Zero and put six or seven second burst into him. He began to pull up and lost his starboard wing about halfway out from the wing root and fluttered into the water. The other two peeled off and went back. The Corsair ducked into the clouds and I lost him.

(...) Then I spotted a Betty low on the water and heading about 130 degrees. I thought he had only one Zero for cover. When the Zero got out of position I went for the Betty, firing for the motors. I was about fifty yards behind the Betty when I hit him. He nosed over and hit the water. I pulled up over the splash, made a turn and headed for the clouds to get out of the way of the Zero, but found another making a pass at me. He made a high side run on me when I was at about 1,000 feet. He got my motor right away – I could see the holes appearing in the cowling. I got into a cloud and in the middle of it the engine began sputtering. I found I had neither oil nor fuel pressure.

SECRET

ACTION REPORT OF JUNE 3o, 1943 CONTINUED VMF-221.

High flying US fighters engaged the Zeros while VMF 221's Corsairs gained a position of advantage from which to attack the torpedo planes, "There they go" was coming over the air and VEGA (Rendova Control) said "Go get 'em, Boys, protect your shipping".All divisions went after the bombers. Twenty seven Bettys in a Vee of Vees began to get Hell. When the fight was over some sections came direct back to base, other planes climbing as requested by VEGA and joining up on others of similar type. These circled shipping for a time and then proceeded back to base, gas getting low. All of VMF 221's planes were back on the deck at 1735, a little ahead of the printed schedule, but with a job well done.

Individual Pilot's scores are as follows:

Capt. Payne	2 Bettys
Lt. Langston	1 Betty
Lt. Baldwin	1 Betty 1 Zero
Lt. Moore	1 Betty plus 1 probable Betty
Lt. Walker	1 Zero plus 1 probable Zero.
Capt. Swett	2 Bettys ½ Zero (assisted by F4F)
Lt. Segal	1 Betty plus 1 probable Betty 1 Zero
Lt. Dillow	2 Bettys plus 1 probable Betty and 1 probable Zero
Lt. Kellogg	1 Betty
Lt. Schocker	1 probable Betty
Lt. Chapman	2 Bettys
Lt. Sage	½ Betty (Assisted by F4F)

The total score for VMF-221:

Enemy shot down	Probables	Own Losses
13 Bettys 3 Zeros ½ Betty ½ Zero	4 Bettys 2 Zeros	Not Any

All of our planes returned safely with but a few battle scars, Our only casualty was Lt. Wood who got a grazing wound in his left knee while scrapping with a Zero. The pilot's individual reports are as follows.

CAPT. J.S. PAYNE " On tally-ho saw twenty seven bombers,in a large Vee formation, down low, headed for the shipping below. I dove on the formation and shot down a Betty, then pulled up, looking back, and saw another with it's port engine burning but heading direct for a ship. I dove on it, firing on its starboard engine, which burned and caused the entire plane to appear to be flaming. I pulled up and looked back - there was no bomber".

③

- 2 -

SECRET 2

"Go get 'em, Boys"... VMF-221 scores big on the day of Rendova landings (action report of 30th June 1943).

Corsairs of VMF-222 on the ground ready to taxi onto the runway for launch from Russell Islands, July 1943.

We went down immediately and broke out of the clouds into the fight at about 6,000 feet over Blanche Channel. I saw a Jap zero dead ahead on the tail of a F4F and pulled behind the zero and shot him down. Immediately on my left another F4F went by with a zero getting into position on his tail. I shot that zero down, as I did the first. Both zeros went down burning. A third zero started a head-on pass at me and I started to turn into him. Just before I could bear on him, he opened fire and hit my left wing stub, coolers, the underside of the engine, and the left side of the fuselage, forward of the cockpit, and my plane commenced to burn and smoke heavily. Just after my plane was hit I got my sights on him and got in a good burst and he passed under my left wing, burning. I was in the fight no more than 60 seconds, I figure. I then attempted to get out, but descended to under a thousand feet before I cleared the plane. The tail hit me in the chest when I jumped. The parachute functioned perfectly but I was so low that the chute opened fully when I was only 100 feet above the water". About 15 minutes later Maj. Weissenberger was picked up by a boat from destroyer USS Talbot and delivered to Guadalcanal next day.

Unlike the Zeros, biplane floatplanes were no match for the Corsairs. Lt. Sheldon Hall, who got two, reported: "About 1800 we sighted bogeys coming in from the SW. There were 9 float biplanes at 10,000 feet. We were the only four planes to make contact with these 9. They were strung out in no particular formation. I dove at the rear plane. Shaw was on my wing. I fired at close range and saw the shots go through him stem to stern. He blew up in the sky. I kept going looking for another target. Shaw left me then to pick his own target. We were all four pretty much together all the time. I saw another float plane cross in front of me heading toward Rendova. He dove away and I dove after him. I fired on him at close range from the rear. I saw the rear gunner killed, and the Jap plane went completely out of control, spinning into the mountains on Rendova".

VMF-221 intercepted a formation of Bettys in the middle of the their torpedo run on invasion fleet off Rendova, scoring the first major success since the squadron's conversion to Corsairs – 16 individual victories (13 Bettys and 3 Zeros) and two more (Betty and Zero apiece) shared with Wildcat pilots. VMF-121 pilots were credited with 3 Bettys and 15 Zeros; Capt. Robert Baker and Capt. Perry Shuman got three apiece. VMF-122 contributed with four victories: a Zero and three Bettys. The bombers no doubt belonged to a mixed formation of 17 Bettys from 702nd Kokutai and 9 of 705th Kokutai. In all, 19 out of 26 bombers failed to return; their only success was a single torpedo hit against a transport ship.

By the day's end the AirSols pilots fighting off Japanese attacks on the invasion fleet and the beachhead were credited with a total of 100 victories.[4] Of this number, Corsair pilots racked up 58. The AirSols lost 14 fighters and seven pilots, including 11 Corsairs and five pilots killed or missing in action.

In early afternoon of **2nd July**, shortly after patrolling fighters had been recalled from above Rendova due to weather front closing in, the Japanese carried out a surprise bombing raid. Some 18 Bettys escorted by 20 Zeros hit the island, killing 59 allied servicemen on the ground. AirSols Corsairs hastily returned over Rendova and before long VMF-121 pilots shot down six Zeros, two apiece by Capt. Robert Baker and Capt. Perry Shuman.

On **7th July** 12 bombers protected by 60 Zeros attempted to attack the invasion fleet off Rendova. They were intercepted by pilots of VMF-121, -122 and -221, who jointly accounted for 6 Bettys and 10 Zeros. In VMF-122, Maj. Bailey and Capt. McCullah each shot down two Bettys, but the squadron lost Lt. Ewing.

On **11th July** two leading aces of VMF-221, Capt. James Swett[5] and his wingman Lt. Harold Segal, audaciously attacked a formation of 12-15 Bettys escorted by a dozen or so Zeros. In a scrap over Kula Gulf (between Kolombangara and New Georgia) Swett shot down Betty and Zero apiece, and Segal three Zeros, before the two pilots were overwhelmed by the opposition.

After the first firing pass at the bombers, during which Segal shot a Zero off Swett's tail, the two got separated and each fought his own battle. Segal reported:

"When I got my altitude back the bombers were a good distance from me and I knew I couldn't catch them. But there were about ten Zeros below and about three miles from me. I saw no friendly planes. While climbing I called Swett but heard nothing. Two of the ten Zeros were off to one side and more or less by themselves. They were about 4,000 feet below me and separated by about 2,000 feet in altitude. I dove down and got the top Zero. It blew up. I continued in the same run, still firing, gave my plane a little left ruder and got the second as well. He began smoking, then flamed and went down. Then I felt bullets going into the front of my aircraft. One came into the cockpit and went through my radio. I could see three Zeros coming from my right, one of which had me perfectly bore-sighted and was shooting the hell out of me. I did a 'split S' and dove down, while diving Zeros were making runs on me from all sides. I almost rammed one. I kept

Lt. Donald Balch of VMF-221 contemplating his luck after he returned in his badly shot-up F4U-1 on 6th July 1943.

Lt. (later Lt.Col.) Kenneth Walsh at the time he served with VMF-124. One of the corsair top-scorers, he tallied 20 victories between February and September 1943.

F4U-1 Corsair piloted by Maj. Robert Owens, the commander of VMF-215, the first US aircraft to land on Munda, New Georgia; August 1943.

Then it quit. I got squared away and headed in a general direction of home, making a water landing about five miles off shore. The ship made a good landing. The hatch stayed open and the plane skipped along nicely. It gave me plenty of time to get out. But there were two Zeros firing bullets into the water all around me so I dove right overboard. I didn't even inflate my Mae West, and left my chute in the plane, getting into the water as fast as I possibly could. The tail of the plane stuck out of the water for about ten minutes and I spent my time ducking around behind it every time the Zeros came by. They were coming in on opposite courses and I was quite active for a while. After about five minutes they left and I was able to inflate my life jacket". Swett swam to Rendova, from where he was picked up the following day – sunburned, but otherwise OK.

During the day Lt. Albert Hacking of VMF-221 was credited with a rare trophy – a Japanese twin-engined fighter, which he shot down northwest of Vella Lavella.

VMF-213 also joined in the fight over Kula Gulf, scoring four victories, but two out of the seven participating pilots, Lts. A.R. Boag and W.J. Thomas (the latter went on to become one of the Corsair top-scoring aces) were shot down; fortunately both survived.

Lt. Boag had a particularly rough experience. After he clobbered one Zero, another got on his tail and shot his Corsair full of holes. Boag reported: "I tried to outturn him, but he stayed right with me, bouncing bullets off my armor plate like hail off a tin roof. I rolled my plane over on its back and dove straight down. I think he followed me down, but when I pulled out at about 10,000 ft he was nowhere

This F4U-1D Corsair named 'The Spirit of 76' was the mount of Maj. Robert Owens, the commander of VMF-215; Munda, August 1943.

around. I was heading towards Munda at a ter-rific speed. I saw another Zero heading straight for me, head-on. I opened fire and only had one gun firing, but I kept firing right into him, knocking pieces off his cowling. He tried to ram ne so I pulled up to the right and as he passed ie exploded in the air. The concussion from the explosion gave my plane a terrific jolt.

I noticed tracers passing me so I started zig-zagging. The tracers and 20 mm were hit-ing my wings near the ailerons. I could see daylight through the holes. I started to roll over nd dive when I saw another Zero coming up from beneath me and forward, shooting all the way. None of my guns would work at this time. I slipped my plane from side to side, still receiv-ing hits on my plane from the rear".

Boag outdived and lost the pursuing Ze-ros, but his ordeal was not over yet. He man-aged to reach a point off of Segi, where he at-tempted to bail out. "My engine cut out again. I was about 6,000 ft and put my plane into a loop and on reaching the top of the loop I at-tempted to get out but something was holding me. I kept my foot on the stick, finally getting partially out, and my parachute pack caught on

This F4U-1 #7, carrying the name of 'My Bonnie', belonged to VMF-124; Munda, August 1943.

Capt. (later Lt.Col.) Arthur Conant served with VMF-215, scoring three victories in August 1943, and another three over Rabaul in January 1944. During the Korean War he flew F9F-2Bs in close air support.

something on the canopy. I was able to keep one foot in the cockpit and keep the stick forward, otherwise the plane would have gone into a dive. I pulled myself into the plane, my parachute was freed of whatever it was caught to and I was able to get out. The aerial struck my hand severing the ends off of the third and fourth fingers on my right hand, also the tail hit me. When I knew that I was free of the plane, I pulled the ripcord, the parachute opened with a sudden jerk and then let me down, slowly, into the water".

Apparently, the air battle over Kula Gulf was again a case of too many aircraft types at the same time in the same piece of the sky. Maj. Gregory Weissenberger, who got one Zero, reported: "I had a great deal of difficulty in distinguishing between the P-40s and the Zeros. In one instance I dove on a Zero, but on close observation I saw two white stripes abaft of the cockpit and withheld my fire, later to find out it was a Zero. Another time I dove on a P-40, thinking it a Zero and was on the verge of firing when I realized it was a P-40".

In the morning of **12th July** Capt. Herbert Long of VMF-122 'made ace' by shooting down a single Zero over Rendova.

A curious remark appears in VMF-213 diary on that day: "During the course of the patrol, Vega [ground controller] reported a flight of enemy bombers and fighters coming in over the North of Kolombangara. The F4Us were kept on station and told to be prepared to intercept. About 12 Zeros were observed, apparently in a dogfight over Vila, but Vega called the patrol and told them to stay on station, that the zeros were staging a dogfight among themselves in hopes of luring the F4Us off of their patrol and thereby allowing their bombers to get in to their objective. The F4Us stayed on their patrol and made no contact". Three days later *Hellhawks* did make contact and scored heavily.

On **15th July** two Corsair squadrons completely broke up a raid by Betty bombers and some 50 Zeros. First to attack were eight VMF-122 pilots led by Maj. Joseph Reinburg. Vectored by a radar operator aboard a Navy destroyer, they made contact near Rendova. On seeing the Corsairs, Betty crews began to jettison bombs and turn back. After the initial skirmish Reinburg found himself alone with 13 bombers scurrying away at full throttle in a shallow dive. He gave chase and splashed two before a lone Zero caught up with him and drove him away. It was a big day for VMF-122 – Reinburg and his seven

F4U-1 #61 of VMF-215 at Munda airstrip in August 1943.

subordinates racked up 14 victories. Lt. Paul Fuss shot down a Betty and two Zeros; Capt. Ernest Powell went one better – two Zeros and two Bettys.

The bombers hassled by Maj. Reinburg didn't get far. Near Vella Lavella they were intercepted by eight pilots of VMF-213, who piled up 16 victories: 10 Bettys and six Zeros by the day's end (against the loss of one pilot). Among the victors were Capt. James Cupp (Betty and Zero apiece), Lt. Edward Shaw (two Bettys and a Zero), Lt. Wilbur Thomas (two Zeros and a Betty), Lt. Milton Vedder (three Bettys), Lt. Sheldon Hall (Betty and a Zero) and Lt. John Morgan (a Betty and a Zero).

On **17th July** four Corsair squadrons (VMF-112, -122, -213 and -221) formed part of a 120-strong fighter formation escorting 32 SBDs

Another F4U-1 of VMF-215 at Munda, New Georgia; August 1943.

F4U-1 Corsair #91 of VMF-124 on the ground at Munda, late August 1943.

and 32 TBFs on a low-altitude strike against shipping in Tonelei Harbor and in Kahili/Buin area. The enemy put up a spirited defense, and in the ensuing air battle Corsair pilots amassed no less than 41 victories. Amazingly, although several Corsairs returned badly shot-up, none was lost.

VMF-122 notched up 14 victories (twelve Zeros and two float planes); Capt. Martin Lundin shot down two Zeros by himself and shared a third with another pilot. In VMF-213, which also racked up 14 victories (twelve Zeros, one float monoplane, and one float biplane), Lt. John Morgan and MG. (Marine Gunner) Gordon Hodde each shot down three Zeros, while Lt. Edward Shaw accounted for one Zero, one float monoplane and one float biplane. Lt. Albert Hacking of VMF-221 went one better, downing four Zeros over Moila Point on Bougainville. In VMF-112, Lt. Ewing got two Zeros. His squadron mate, Lt. Bougreois, was about to knock down another Zero, when three P-40s opened up on him. Bougreois dropped his wing to show the white star and they ceased firing, but the smoking Zero slipped away.

On **18th July** 56 bombers (B-24s, SBDs and TBFs), escorted by 134 AirSols fighters, struck at Kahili. Corsair pilots accounted for 10 Zeros. Nonetheless, it was a tough day, for five Corsairs failed to return, including two flown by aces. Capt. Ernest Powell of VMF-122 was posted missing in action, presumed killed. Lt. Sheldon Hall of VMF-213 returned to the squadron in early August, with quite a story to tell:

"B-24's approached the airfield from the North. The AA fire was very plentiful as were the zeros and we were kept busy chasing the zeros away from the bombers. As the zeros came in on us, Shaw and I got one each, both of which I saw go down, burning.

About this time we saw a Corsair about 3,000 feet below us with 3 or 4 zeros on his tail, so we went down to help him, because he was, definitely, in trouble. We took the zeros off his tail, each of us tallying one apiece. It was a fight all the time. We really worked like a team. At 5,000 feet a zero put a 20mm through my throttle quadrant, wrecking it completely. Also I got some shrapnel in my left hand and wrist.

I didn't see Shaw after that but I called him over the radio and told him to go home. I doubt seriously if he heard me. My engine had quit as soon as the 20mm hit. I was afraid to bail out for fear the Japs would shoot me. I was also afraid to slow down the plane for fear he would get somewhat of a beam shot at me and then I would be a dead duck. I was sure mad as hell though, cause I couldn't do anything.

All the way down to the water from 5,000 feet the Jap just sat behind me and shot. The bullets hit my armor plate like hail hitting a washboard. My wings and engine were terribly shot up. I knew I was getting close to the water so I jettisoned my cockpit hood and got ready to hit. Using no flaps or anything I landed at 150 knots indicated. I remember the first skip of the plane and then all went black.

The next thing I knew I was hanging onto my inflated rubber boat (seat type), my life jacket was inflated and my parachute was in my boat. How I got out of the plane and did all this will always be a mystery to me.

About this time I saw the zeros coming at me so I shoved myself away from the boat and put my head under water, as the Mae West wouldn't permit the rest of my body to submerge. I heard the zero pass over and came up to look at him. He did a wing over and came right back, as I carried out the same procedure and for the second time he missed both myself and the rubber boat. He left me then so I crawled into my boat.

Everything seemed to be pretty well intact about me, except that my face was a mass of blood from a broken nose and several front teeth missing. I had quite a few shrapnel wounds on my left hand and wrist and 7.7mm skin creases here and there over my body, but I could see and hear and no limbs were broken so I figured that I wasn't too bad off".

Capt. James Cupp of VMF-213, an ace with 12 victories scored between mid-July and mid-September 1943.

Hall paddled towards the nearest island in sight, which happened to be Choiseul. After a few days of wandering around he ran into friendly natives, who led him to their village, and eventually to a spot from where he was picked up by a PBY flying boat.

Meanwhile VMF-221 was struggling with mechanical problems, which almost grounded the entire squadron. On 18th July squadron diarist noted: "Our oil trouble, asphalt like sludge in the sump strainers, gets worse", and the following day: "Planes available only for local emergency scrambles because of the oil trouble". During one of the local hops Lt. Baldwin strafed a shark, which he mistook for a submarine.

Fresh Corsair forces were arriving. On **22nd July** VMF-214 *Swashbucklers* relieved VMF-121, which returned to the USA. On **25th July** VMF-122 was relieved by VMF-215 *Fighting Corsairs*, the future top-ranking Marine squadron in the Solomons. Three days later VMF-124 returned for yet another tour of duty (relieving VMF-213).

On **6th August** the *Swashbucklers* (VMF-214) and VMF-221, eight Corsairs from each squadron, were tasked with escorting an F-5A (a photo-recce Lightning) to the Shortlands. Their sudden arrival triggered quite a commotion at a local anchorage of Japanese seaplanes. Several A6M2-N Rufes (a Zero floatplane variant) and E13A Jakes scrambled. Before long they were joined by Zeros stationed at nearby Ballale. The *Swashbucklers* stood their ground, claiming five for the loss of one pilot; 2/Lt. Alvin Jensen clobbered three – two Zeros and a Jake over Kulitanai Bay. VMF-211 got three, including two by Lt. Harold Segal, who reported:

F4U-1 Corsair of VMF-214 at Espiritu Santo, September 1943.

"I saw a plane pull around in front of Gene [Lt. Eugene Dillow] and he went up after him. Another dove between us and got on his tail. I saw the Zero shooting at him, but I was right behind the two of them and was afraid to fire for fear of hitting Gene. I called him and told him to get out or he would get the Hell shot out of him. He did a roll and went straight down and left a clear shot at the Zero. I poured it into him until I saw smoke and flame. Then I felt bullets going through my wing and all of a sudden the plane shook violently and I saw that half my left aileron had been shot away by a 20 mm. I saw I couldn't make a tight enough turn to follow Gene so I dove out and tried to make for a cloud. I wasn't looking for any more fighting but as I headed for the cloud I met a Zero coming at me head on. I would have let him alone but he was dead ahead so I pulled a little to the left and gave him a good burst. He began to smoke and as he passed over me I looked up and saw flames shooting out from under his cockpit. Then I got into the cloud, bullets following me all the time."

The following day VMF-214 lost its commander. Maj. William Pace took off on a test hop of his assigned plane in which a new engine had been installed. Coming in to the traffic circle his engine failed. Unable to land as the runway was at that time occupied, and also evidently too low to turn back to the water, he either jumped or attempted to have his parachute pull him off the plane by standing in the seat and pulling the cord. In any event, he was too low to the ground, and he was killed.

On **12th August** Lt. Kenneth Walsh became a 'double ace' with 10 victories, but was lucky to survive. He reported:

"Our Squadron (VMF-124) composed of 16 Corsairs and led by Maj. Millington took off from Fighter Strip #1 at 1015. We rendezvoused at Savo Island and proceeded to the Russells to further rendezvous with the bombers. While joining up on the B-24's Major Millington left the formation due to engine trouble and landed at the Russells. This gave me the lead of all fighters on the mission, my radio call being 'Golden One'.

(...) In the vicinity of Choiseul Bay, about 30 zeros found us and immediately pressed home their attack; at first striking from below, head-on, then from the rear and side. When attacking head-on they would roll over on their backs after firing their guns. About five minutes after we were first attacked, while letting down between a valley of clouds, so to speak, half of our fighters got separated from the bombers and were unable to find us. This was most unfortunate since we had few enough fighters as it was, and this left only 13 fighters to cover the bombers in a running fight for about 100 miles. The 13 fighters left were 5 Corsairs and 8 P-40s, the Corsair pilots being my entire flight: Lt. Johnston, Capt. Raymond, Lt. Willcox and myself and Capt. Aldrich of VMF-215.

Needless to say the fight got pretty rugged, especially for the bombers in the rear of the formation since the zeros were paying particular attention to them. Sometimes a zero would just hang for a while about 500 yards astern, fire, roll over on his back and pull away – it was obvious they were after tail gunners in the B-24's. To counter-attack this, especially since nothing was coming in on us from above or forward, I decided to make overhead runs on what zeros that would venture in the closest. Johnston (my wingman) and I would maintain about 5,000 ft. above the center of the formation, wait until some bird came in close enough, then down we'd go. This procedure worked beautifully for 4 or 5 passes, in which time I am certain I destroyed two zeros of the Zeke type and probably a third. I had my camera gun loaded for this flight and the film should confirm the above score. The first overhead pass I made was the best of any I have ever made including target shooting on a sleeve. The Jap never saw me, I opened up early and followed through with a long burst which lasted until I was so close I had to duck from colliding. While I was firing he rolled over on his back and white smoke seemed to be coming from his belly; looking down after I passed I observed him in flames. Since the bombers had to slow down in order to maintain a tight formation it was easy for us to make the overhead runs with enough speed to always pull up through the bombers to our original position. But as I said before it worked for 4 or 5 runs.

Apparently one of the zeros got wise to my game and as I started down on my last overhead (which was really a high side this time), a zero got on my tail which I never saw. Johnston said [the zero] got so close on my tail that he had to delay shooting him off my tail for fear of shooting me as well. This is obviously true since in a very short burst I was hit with 7 cannon shells and over twenty 7.7".

Johnston reported: "After about our fourth pass Walsh made a high above side run and a zero came streaking down on him just as he pointed his nose down. I cut to my right and pointed the nose of my plane at Walsh and as soon as his rudder fin passed I opened fire and the zero who was firing at Walsh went through my fire, rolled over and disappeared beneath me. I did not watch him as I was more concerned with Walsh whose plane was streaming smoke."

Again Walsh: "As soon as I got hit my cockpit filled with smoke. I immediately rolled over on my back and started to bail out. I knew I was hit bad; it sounded not like rain but rocks on a tin roof. However, when I opened my cockpit cover, or greenhouse as they now call it, the smoke dissipated and I discovered its source below my seat. I decided not to bail out until I saw flames. I can't say how grateful I am for having a wingman like Johnston – another second later and I surely would of 'got it'. He not only shot him off

F4U-1 Corsair of VMF-214 at Turtle Bay fighter strip on Espiritu Santo, September 1943.

2/Lt. Henry Huidekoper of VMF-213 and his F4U-1 Corsair named 'George' after getting shot up by a Zero pilot over Kahili on 27th September 1943.

my tail but shot him down as well. I immediately pulled up under the bombers for protection and remained there until we were opposite Segi.

(...) Due to my hydraulic system being completely shot out I had to land without flaps or brakes at Segi. I managed to get the wheels down by using the emergency system but after I landed and rolled about three-fourths of the way down the field, the plane went completely out of control thus causing me to crash into a plane parked near the runway."

Capt. Donald Aldrich of VMF-215, whom Walsh mentioned in his report, scored his first victory (a Zeke near Ballale) on that day. By the time his third combat tour in the Solomons was over, he had 20 to his credit.

The following day VMF-221 left for Sydney, relieved by VMF-123 *Flying Eight-Balls*. Around that time the captured and rebuilt airstrip at Munda was put back on operations. Immediately VMF-123 and -124 took up residence there. The timing was perfect, for the following day

F4U-1 Corsair of VMF-223 at Munda, 1943.

F4U-1A Corsairs of VF-17 aboard USS Bunker Hill passing through the Panama Canal, September 1943. Shortly afterwards, upon arrival at Pearl Harbor, the squadron was ordered to leave the ship and move to the South Pacific to fight from land bases.

the Marines again surged forward. They by-passed the heavily defended Kolombangara and landed at Vella Lavella.

On **15th August**, the day of Vella Lavella landings, VMF-124 scored the biggest success in the squadron's history, tallying 13 victories. Lt. Kenneth Walsh shot down a Zero and two Vals. Doubles were credited to Lt. Howard Finn (two Vals), Capt. William Cannon (two Zekes) and 2/Lt. Troy Shelton (two Zekes). Finn's report read:

"Crowe, Nichols, Mayberry (VMF-123) and myself were patrolling over Cracker Base at 10,000 feet. A bogie was reported. We picked up four Vals and started diving down. Crowe made a run on the first dive bomber and the one following turned on Crowe's tail as he went by. I then came in and started firing on the one that was following Crowe. It took quite a burst before it finally exploded.

I slowed up and set down behind another dive bomber and started firing. I fired, fired and fired but the damn thing kept going straight ahead. The rear gunner was dropped over the side like a wet rag. It was at this time that stuff started coming thru my wings. One bullet came into the cockpit hitting me in the hand. I then looked and saw six zeros behind me so I gave it full throttle. An F4U was coming from the opposite side. It turned out to be Crowe making another run and I headed over to join up with him. We scissored perfectly and I started a circle to the left and Crowe one to the right. One zero went in after we scissored. I suppose Crowe got him.

(...) I was shot up pretty badly and upon landing the plane swerved to the right. I could not bring it back on the runway, hit a shell hole and went over on my back. When I got out of the hospital I had two planes to my credit so the dive bomber I filled full of enough lead to sink a ship must have gone in."

For VMF-123 the day was just as intense. The squadron's war diary relates:

"Patrols and scrambles were made throughout the day. Contact with enemy planes were made in the morning and afternoon around Vella Lavella while covering Task Force and landings on this island. While flying at 21,000 ft., at 0750, a three-plane flight of F4U's, consisting of Baker, Bowles and Noonan hit 30 to 40 zeros flying at 20,000 ft by striking out of the sun and at the rear of the formation. Baker knocked

Maj. Gregory Boyington, the commander of VMF-214 *Black Sheep*, at Barakoma Airstrip on Vella Lavella, November 1943.

down 2 of them and Bowles one. Bowles also destroyed a dive bomber a few minutes later, while it was traveling over water at 50 ft."

Indeed it was a perfect bounce – despite the odds! Bowles related:

"We swung left, made a 180° turn, came out of the sun and Major Baker led the flight in on the tail of the formation. Baker hit 1 Zero on the first pass and it blew up in front of him and the Major went right thru the flames. I then got a burst on tail-end Charlie while he was making a steep climbing turn, and made him smoke. I climbed up to his tail and just as the pilot was about to abandon his plane I put another burst into him. The plane blew up."

In this engagement VMF-123 won five victories (a Val and four Zeros) for no losses of its own. In the afternoon two pilots of VMF-123 and three of VMF-124 formed a five-plane flight, which in the vicinity of Vella Lavella bounced a formation of about 40 Zeros escorting 15-20 Vals. The two pilots of VMF-123 each scored one victory, for a total of seven for their squadron.

On **18th August**, another day of the air battle over Vella Lavella, the *Eight-Ballers* were credited with a single victory (a Val) but lost two Corsairs and one pilot. 2/Lt. Lisle Foord just went missing; it was only presumed that he took off on his own, joined a fight and was shot down. 2/Lt. Foster Jessup was hit by friendly AA and forced to bale out. A few days later, while aboard an LST, he volunteered to man a .50 caliber machine gun. Shortly afterwards the ship was dive-bombed by eight Vals, and Jessup had an opportunity to fight from this unique position (and nearly got himself killed – a bomb blast from a near-miss knocked him unconscious). The Navy credited him with probable destruction of two of the attacking aircraft.

On the same day Maj. Raynold Tomes and Lt. Harold Langstaff of VMF-215 were sent on a reconnaissance mission to observe shipping at Kahili Harbor from high altitude. Steadily climbing, they arrived midway between Kahili and Kieta airfields at 32,000 feet. While counting ships, they spotted six Zeros in formation at 17,000 feet, and in one pass shot down two of them.

On **19th August** four Corsairs of VMF-215 ventured over Rekata Bay at the coast of Santa Isabel to hunt for Japanese seaplanes. Lt. Harold Spears, the future 'triple ace' of VMF-215, shot down two floatplanes and destroyed one in water. He reported:

„As we neared the bay, very heavy AA opened up and I saw a float Zero taxing out from shore. I made a run towards the shore on him, and set him afire with a long burst. I spotted some float Zeros just as they had taken off the water. I made a stern run on them, setting both on fire in the same run, as one followed the other. They crashed, flaming, into the bay a short distance ahead on the same bearing as their takeoff course. I then turned back for another strafing run on some anchored float Zeros. I saw smoke but could not observe results as I was busy helping Lt. Sanders at this time". Lt. George Sanders was shot down by AA fire and ditched nearby. He managed to paddle in his raft to the shore before the Japanese could find him, and after a few days he ran into friendly natives, who sent a word to a local coastwatcher.

On **21st August**, during the day's heavy fighting over Barakoma airfield on Vella Lavella, Corsair pilots shot down three Vals and ten Zeros. Capt. William Crowe of VMF-124, who bagged two Vals and a Zero, reported:

"I was at 12,000 feet and climbing West toward Ganongga. I looked over to the Northwest and picked up a large formation of planes. I knew then that they intended to hit Barakoma. They were at about 15 to 16 thousand feet, and above them there were two groups of fighters. I would estimate there to have been at least 25 dive bombers and 30 fighters in this first wave.

My wingman, Lt. Mutz, was right on my wing, but the rest of the flight was strung out behind. The Zeros spotted us almost immediately and swung over and started to get in position from above on our tails. I had headed for the dive bombers before this but was still below them. I had boosted my RPM-throttle, but not full because I wanted the boys in the rear to close up on me. But as soon as I saw the Zeros head over to get on our tails, I gave it full everything and started climbing at 175 knots and 2,500 to 3,000 feet per minute (Believe it or not!). Lt. Mutz stayed right with me, but the other four boys were left behind and the Zeros started attacking them – also closing on us from the rear. It was a pretty close race to see whether we were going to get in position for a run on the dive bombers or whether the zeros would get to shooting range on us (Lt. Mutz and myself).

We won! We got in position for a good high side attack and made it. I got the trailing Val on this pass, pulled up in a wing-over and made another high side on the leading bomber, but missed. Just at this time I saw another group of dive bombers and Zeros coming from the North – equally as large as this group. By this time the lead bombers had gone into their dive and I immediately tailed in on one of the two leading bombers. I closed and started shooting at about

Corsairs, Hellcat, Dauntless, and New Zealand Kittyhawks at Barakoma on Vella Lavella, early December 1943.

An unidentified F4U Corsair touches down at Torokina airstrip on Bougainville, late 1943.

8,500 feet and followed him to about 4,500 feet where he started his pull up and blew up.

(...) By this time Zeros were trying to make passes on me from every angle. I had lost Lt. Mutz by this time and felt pretty lonely as far as friends go. I had plenty of Zeros for company though. So I nosed down (still full everything) and caught a Zero trying a beam attack. I pushed around and my 4 – 50's beat him to the punch. He blew up (I only had four of my six guns working, two on each wing). I straightened out and saw two Zeros directly ahead and started going for them. I got a few bursts at one of them, but didn't see anything happen. I looked up and saw they were taking me toward the Northeastern tip of Vella Lavella where there was a whole flock of Zeros just milling around. I immediately turned and headed for the coast of Kolombangara. Four Zeros gave chase and stayed right behind me until I reached the Eastern side of Kolombangara in Kula Gulf. Those boys didn't gain any on me, but they sure didn't lose any ground – and I was plenty scared – because I was by myself – with 40 gallons of gas showing on the gauge. I didn't know when they were going to turn back or catch me. Well they finally turned back and I immediately eased off on my throttle and RPM for the first time since putting it on for my climb into the dive bombers. I returned to the field and landed – and considered myself damned lucky. Lt. Mutz landed just a little after me pretty well shot up and with just about the same experiences I had."

Lt. Thomas Mutz shot down one Val before he, too, was jumped by the Zeros. His report is a testimony to how much beating the F4U could take:

"They knocked out my right wing guns, one-third of my right aileron and most of my right wing flap. I dove away and when I pulled out going North along the coast of Vella Lavella with Zeros chasing me, I made a head on with a Zero. I could not see the effect of my shooting and I had to correct for the fact of having only three left wing guns working. I then pulled away from them and throttled back as I had on full power ever since climbing into the fight and my red carburetor temperature warning light had been on most of the time. I was then up to the Northern Coast of Vella Lavella and my gas was getting low. Then a Zero which I didn't see at the moment made a low side run on me putting many rounds in my engine and prop. A 20 mm exploded inside the cowling knocking out the oil sump and several cylinders and other general damage. At this time 4 more Zeros dived on me from above-left and I got a good shot at one of them. By this time I was wide open again and down on the water. Luckily they just made this one pass at me because they could have easily finished me off".

VMF-123, the other Corsair outfit stationed at Munda, was equally busy throughout the day. Before noon the *Eight-Ballers* scored three Zeros but lost a pilot; in the afternoon they got two more Zeros. It was, as usually, no quarter asked and none given. One of the pilots reported:

"Having heard some dope on the radio about Vals rendezvousing N. of Baga, I went there as fast as possible but found nothing. I started home and saw a Val in the water about 100 ft from the west shoreline of Baga. I turned and strafed it. I am fairly sure I killed the pilot who was running out to the wing tip when I fired. He was motionless in the water when I came back to take a look."

Meanwhile, the future 'top guns' of VMF-215 were gaining their first victories and some badly needed experience. The squadron had sent one flight under Maj. Robert Owens up to Munda to strengthen the local forces. While on station the flight was ordered to Ganongga Island to help intercept a large force of about 90 Japanese planes. Maj. Owens spotted five or six Zeros below and led his flight down on them.

Maj. Owens: „I picked a Zero while in my dive and came up astern of him. I fired a ten second burst into him but nothing seemed to happen. I then made a 90° turn to the right and came up on another Zero in a stern run. I put a good burst into this one and he rolled off and went straight down, flaming and smoking".

Lt. Arthur Conant: „I was diving with Major Owens, following him at a slight distance, and saw him shoot at his first Zero. The Zero flipped over in a left split S, then dove. I followed him in his maneuver and was just closing in on him when tracers which might have been Lt. Newhall's went past me. Thinking I was being fired upon, I took evasive action, then tried to get back on the Zero. I was going too fast and passed him however. I pulled out of this dive, blacking myself out. When I came to I was alone".

On **23rd August**, another day of the air battle over Vella Lavella, VMF-124 tallied five Zeros (including two credited to Ken Walsh). VMF-123 pilots shot down four Ki-61 Tony fighters of the Imperial Japanese Army Air Force, never before reported by AirSols pilots. One of the victors observed: „There were twelve in all – in-line engines, square wing-tips, silver prop spinners, red meat-balls without white background and the plane was brown olive drab color. It looked like a P-51 except it was stubbier."

On **24th August** VMF-123 scored big, tallying two Vals and seven Zeros in a scrap over Kolombangara. Lt. Walter Mayberry, who got three Zeros in quick succession, reported:

"While searching for enemy aircraft over Vella Lavella, Lt. Noonan and I saw 16 Zeros (Zekes) ahead and above us. We were at 12,000 ft. when we saw them going N.E. The Zeros were at 14,000-15,000 ft. We ducked into a cloud and climbed to this level. As we came out (15,000 ft) the Zeros were slightly below us and in a right hand turn in groups of four. We made a pass at the trailing group and I shot down in flames the number 2 man. I then did a half split S in order to reach the leading 4 plane group. They were about 500 ft. below the trailing group. I took the trailing plane and as I came within range fired a burst. He immediately blew up. The remaining three planes scattered, one diving straight ahead. I tailed him and blew off part of the right wing at about 13,000 ft. The pilot jumped. I circled the parachute but didn't attempt to kill him while he was in the chute. I looked around. All enemy planes were gone. I found a F4U, joined up and came home."

Lt. Noonan, who clobbered one, related: "The Zero I shot tried to turn into me but was unable to make the turn, consequently I had a good shot at his cockpit. He smoked, rolled on his back and dove into the water."

It tells a lot about the trust Corsair pilots must have had in their mounts – the two lieutenants, none of them a high-scoring ace, stalked the 16-strong enemy fighter formation, attacked it and, with apparent impunity, wiped out one fourth of it.

On **25th August** VMF-215 scored four victories for no losses of their own. The squadron was ordered to escort B-24s to Kahili, but the formation scattered in towering clouds, and eventually only nine Corsairs, four P-40s and six B-24s made it up to the target. As the Corsair pilots broke out of the clouds, they saw below 40 to 50 Japanese fighters closing in on the bombers. Maj. Owens rushed to the rescue, quickly bagging one Zero. Meanwhile his wingman, Lt. Conant, was very busy defending Major's tail:

"I was behind Major Owens shooting at Zeros that were making runs on him. I saw one Zero to the side and made a stern run on him. He fell off and I hit him again. He spun down out of control and hit the water. I came up be-

F4U-1As of VMF-216 at Torokina, Bougainville.

One of VMF-214 Corsairs at Bougainville, 1943.

hind another Zero and poured a long burst into him. There was another Zero on Major Owens' tail on which I made a 45° stern run. I saw my tracers pouring into him and he seemed to stop still. About two yards of his wing fell off among other pieces from his plane. He fell spinning and out of control into the water." These were Conant's first two confirmed victories (he was also credited with two 'probables'). As a matter of fact, his two 'sure' victories were a Zeke (A6M Zero) and Tony (Ki-61 Hien) – 'Zero' often being used by American fighter pilots as a generic name for all Japanese fighters.

In the afternoon of **26th August** VMF-214 and -215 escorted B-24s to Kahili. Their adversaries were again a mixed bag of Tonys and Zekes. Corsair pilots tallied 11 victories, including three by Lt. Hartwell Scarborough of VMF-214. One of the Zero pilots they fought with was Hideo Watanabe of 204th Kokutai, an ace with 16 victories, who was severely injured (he lost an eye) in the scrap with Corsairs.

VMF-215, getting deadlier with every passing day, scored six, including two by Maj. Raynold Tomes. Capt. Donald Aldrich shot down one Zero (for his third confirmed victory) and was lining up another, when he nearly got clobbered himself: "Before I could get into range, something jumped me from behind and a 20 mm shell knocked off some of my rear control surfaces. The rudder was hit badly and I got several small fragments in my right arm and near my right eye".

On **30th August** B-24s again set out to strike Kahili. The escorting Corsairs shot down 16 Zeros, of which 7 were credited to VMF-124. A week later VMF-124 left for the USA, where Kenneth Walsh was awarded, as the first Corsair pilot, the Congressional Medal of Honor. The citation read:

"For extraordinary heroism and intrepidity above and beyond the call of duty as a pilot in Marine Fighting Squadron 124 in aerial combat against enemy Japanese forces in the Solomon Islands area. Determined to thwart the enemy's attempt to bomb Allied ground forces and shipping at Vella Lavella on 15 August 1943, 1st Lt. Walsh repeatedly dived his plane into an enemy formation outnumbering his own division 6 to 1 and, although his plane was hit numerous times, shot down 2 Japanese dive bombers and 1 fighter. After developing engine trouble on 30 August during a vital escort mission, 1st Lt. Walsh landed his mechanically disabled plane at Munda, quickly replaced it with another, and proceeded to rejoin his flight over Kahili. Separated from his escort group when he encountered approximately 50 Japanese Zeros, he unhesitatingly attacked, striking with relentless fury in his lone battle against a powerful force. He destroyed 4 hostile fighters before cannon shellfire forced him to make a dead-stick landing off Vella Lavella where he was later picked up. His valiant leadership and his daring skill as a flier served as a source of confidence and inspiration to his fellow pilots and reflect the highest credit upon the U.S. Naval Service."

For VMF-123 *Eight-Balls*, also nearing the end of their tour in the Solomons, 30th August was the last day of heavy fighting. The squadron's diary related:

"Twelve squadron pilots, with approximately 32 other fighters, escorted 26 B-24s to strike Kahili. Twenty-five to thirty five Jap fighters attacked the formation just as the bombing run was being made, and AA fire was accurate and heavy. A Zero got on Mayberry's tail. Major Baker shot at this Zero and saw him peel off and start down but did not see him crash. The Zero apparently did substantial damage to Mayberry's plane for he lost altitude and went down off Fauro Island. (...) No reports or information was received concerning Mayberry, who is missing in action."

Capt. Jack Scott put up a gallant fight. He reported: "About forty miles southeast of Kahili, one bomber broke up and spun into the ocean with one engine on fire. A B-24 pilot radioed that he was returning to drop him a life raft. Due to the large swarm of Zeros (15) buzzing around, this was a hazardous trip, so I turned to help him out. My flight was jumped from the same altitude (14,000 ft.) and everyone dived out – towards the bombers. Most of the Zeros (10-12 Zekes) were picking on the B-24 who had returned to drop the raft, and who by now was very close to the water (50 ft.).

VMF-216 Corsairs on Bougainville.

A VMF-216 Corsair at Torokina airstrip on Bougainville, December 1943.

I pushed over on through the highest Zekes, shooting at everything which wandered across my sights. On my way down I fired at 4 Zeros and one went down in flames. By this time the B-24 had apparently dropped the life raft to the bomber crew in the water and was heading SW on the water with four Zeros making gunnery runs on him. I saw one Zeke hit the water and another smoking very badly. I trailed in on the rear two, shooting madly with the two guns I had left. They pulled off the B-24 undamaged. The Zeros left the B-24 and I stayed with him until he reached the southern tip of Ganongga Island."

Unfortunately, in postwar years such pilots as Scott or Mayberry (both credited with four victories) rarely got as much attention as the so-called aces, and their exploits are largely forgotten.

It happened in every theater of war that pilots were seeing (and, occasionally, shooting down) planes they couldn't recognize. On **30th August** such thing happened to Lt. Crafton Stidger of VMF-215, who was vectored to intercept a lone 'bogey'. Stidger jumped it over Santa Isabel and shot it down in flames. As the squadron's diary relates:

"During the engagement, Lt. G.S. Stidger had ample time to study the features of the enemy plane. It was painted a dark brown color with red roundels visible on the wings and on the fuselage. The engine was radial with large spinner. The fuselage was bulky, had the appearance of our F6F, with a fused canopy tapering the tail, which was also similar. The wings were tapering with rounded tips. After checking identification composition he is almost certain the plane destroyed was a FW 190 single-seat fighter". And so it was officially recorded – one Focke-Wulf 190 destroyed. Besides this oddity, in August VMF-215 tallied 17 Zekes, 2 float Zeros, 1 Tony and 1 Val – 22 in all.

Around that time VMF-213 *Hellhawks* arrived to relieve VMF-124, starting its third tour in the Solomons. Tough fighting lay ahead for the squadron. In September alone it tallied 28 victories (19 Zeros, 8 Vals and 1 Betty) but lost 10 Corsairs. Five pilots were MIA; five others were rescued, including two aces: Lt. Wilbur Thomas and Capt. James Cupp (the latter, badly burned, was evacuated).

On **2nd September** VMF-215 escorted some B-24s to Kahili. Capt. Donald Aldrich and Lt. Harold Spears each shot down two Zeros. Time and again, Corsair's advantage in speed was proving a life-saving factor; Spears reported: "Four Zeros were shooting at me all the way to Vella Lavella. I got five or six bullet holes in my tail and one in my prop. My gas was low and I was alone. I was able to outrun them back to Munda". Three days later the squadron completed its tour and left for Sydney.

For VMF-213 *Hellhawks* the first major engagement of their third tour came on **11th September**. That day B-24s escorted by some P-39s, P-40s and eight Corsairs of VMF-123, struck at Kahili. Eight VMF-213 Corsairs, along with 16 land-based Navy Hellcats, acted as a delayed fighter sweep. They were instructed to circle over Fauro Island, until the bombers had made their run on Kahili, then to clear the skies of any enemy aircraft that should arise to attack the bombers. The enemy – Zekes and Tonys – came up in force, and the eight *Hellhawks* had their hands full. They shot down seven fighters, including three by Lt. Wilbur Thomas and two by Capt. James Cupp, for the loss of one pilot. As if Zekes and Tonys were not enough, at one moment Cupp's division was fired upon by two F6Fs.

On **13th September** VMF-222 *The Flying Deuces* was twice in action, scoring six victories for the loss of one pilot. In the morning two 4-plane divisions were on a patrol over Vella Lavella, when they were vectored towards eight Bettys escorted by 15-30 Zeros and Tonys. In the ensuing fight the Corsair pilots shot down two Zeros and one Tony but Capt. Moore failed to return. 2/Lt. William Carrell, credited with one confirmed victory after this scrap, reported:

"We all went down together in a good battle formation, and as I pulled out, about 9 or 12 Zeros of the right flank were sitting right in front of me. I opened up on the tail-end man and set him on fire, and then Zeros went in all directions. I got in several very good deflection shots and shattered the greenhouse completely on a Zero with a full deflection shot; he flew right through my guns. I did not see him after that, for there were too many good shots in front of me, but we almost hit – he was so close in front of me.

It was more or less a melee, and I was taking advantage of all shots, all deflection 'till the last one. I opened up on one boy in a 90° deflection, and he turned his tail to me and did a half ass roll. I followed him, and as I got on my back, a lot of dirt, oil, water, etc. from the bottom of the cockpit came up in my face and I couldn't see a thing. I don't know how I got back right side up, but when I got the dirt out of my eyes I was all alone. I pulled up sharp and looked around for my flight or more Japs, but no one was around."

In the afternoon four VMF-222 Corsairs clashed in the same area with 12-18 Zeros. In a rapid exchange of fire they splashed three Zeros (of which two were credited to Maj. A.N. Gordon), then prudently ducked into clouds. Eventually all four returned to base, some with heavy battle damage to prove what they just went through – one pilot by boat, after he had been forced to bail out of his shot-up plane.

On **14th September** 20 Corsairs of VMF-222 escorted a dozen of B-24s to Kahili. In a scrap with some 25 Zeros and 4 Tonys the squadron won five victories, including two by Maj. Donald Sapp. He reported: „My wingman Lt. Jones spotted Zeros below, at 18,000 feet. Not being able to spot the bogey, I turned the lead over to Jones to make the attack. He initiated the attack by diving down out of the sun on the leading plane and overran him after firing at him. He made a hard turn to the right, followed by the two Zeros, one behind the other at about 1,000 feet apart. I closed on the tail of the second zero and gave him a short burst and observed his right wing to disintegrate and the ship went out of control". Sapp continued through to the Zeke closest to Jones and killed the pilot with a burst in the engine and cockpit, observing flames from the engine and the plane going down.

F4U-1A (BuNo 17878) of VMF-214 photographed on Bougainville on 10th December 1943. Note a RNZAF Kittyhawk in the background.

F4U-1A flown by Lt. Bob Marshall of VMF-216 after returning from a dogfight over Rabaul on 19th December 1943.

The same day saw a combat debut of VMF-214 *Black Sheep*. The squadron was formed in a fairly unusual manner. Maj. Gregory Boyington, a veteran of 'Flying Tigers', who had fought the Japanese even before the USA joined the war, was eagerly awaiting an operational posting but none was coming his way. Eventually, securing the go-ahead from his superiors in the 1st Marine Air Wing, he collected 20 spare Corsairs and 27 volunteers from a pool of replacement pilots. The squadron was designated VMF-214, which was the 'recycled' number left over after the *Swashbucklers* had been disbanded. While preparing his subordinates for combat, Boyington instructed:

"You're flying one of the sweetest fighters there is, but there are certain things a Corsair won't do. Don't try to loop with a Zero because Zero is a lighter, more maneuverable plane and will loop inside you and he'll end up on your tail. The same goes for turning – don't try to turn with him. But your ship is faster; it will climb away from him in a shallow climb, and you can outdive anything they've got. So what does all this add up to? Just this: get above him; come in on him in a high stern pass; hold your fire till you're within good, close range; let him have it and watch him burn. When they're hit right they burn like celluloid.

If you miss him, don't stick around to dogfight. Dive out – get the hell out of there – climb away and come back into the fight with some altitude and speed".[6]

The *Black Sheep* drew their first blood two days later, on **16th September**. The assignment was to escort SBDs and TBFs to Ballale. The 24 Corsairs provided top cover at 21,500 feet, with a flight of Hellcats as medium cover at 17,000 ft, and a flight of P-40s at 15,000 feet as close cover to the bombers at 13,000 feet. The defenders scrambled some 40 Zeros. The fight spread all over the sky for 150-200 square miles at all altitudes and continued unabated for 30 minutes. The *Black Sheep* were credited with 11 confirmed victories – not bad for a first fight. Boyington himself, having knocked down five Zeros, became the first Corsair 'ace in a day'. Lts. Don Fisher and John Gebert bagged two Zeros apiece. The mission was marred by the loss of Capt. Ewing, who went missing.

On **18th September** eight VMF-214 pilots participated in warding off an attack by some 50

Vals, which were targeting allied shipping bringing supplies to the Vella Lavella beachhead. The dive bombers arrived protected by 32 Zeros. Lt. Christopher Magee, later to become VMF-214's second ranking ace (after Boyington), daringly dived with the Vals into the Navy's anti-aircraft barrage, and shot down two of them. Moments later he was jumped by the escorting Zeros and badly shot up. He finally eluded his tormentors by slipping under the shield of the naval guns.

During the same engagement eight pilots of VMF-213 shot down two Zeros and eight Vals, but the price to be paid was three Corsairs and two pilots. Maj. Cloake was not observed after going into initial attack. Lt. Glascock's plane received a hit from one of the Vals' tail gunners, his plane rolled over and went into a spin, the left wing burst into flames at its root, and the plane eventually disintegrated and spun on down. As the squadron's diary states: "There is a possibility that the Val which shot down Glascock got a lucky shot into his wing tank which may have not been sufficiently purged."

Slugging it out over Vella Gulf (separating Vella Lavella and Kolombangara), Capt. James Cupp shot down four Vals, his wingman Lt. Walter Stewart three, and they teamed up to bag another Val. Then they were jumped by the bombers' escorts. Cupp got away, but Stewart was shot down and bailed out off the coast of Gizo Island; picked up by natives, he was returned to Munda the following day.

On **20th September** Capt. James Cupp led four Corsairs on a dawn patrol. They were vectored toward a single Betty bomber, which had just bombed Munda. Despite poor visibility they gave chase. Cupp's engine began to act up, so he made a risky straight approach at the Betty from the rear and below. He was hit by return fire and his Corsair burst in flames. According to the squadron's diary, "The plane caught fire immediately in the bottom of the cockpit. When [Cupp] realized his predicament, he turned south and attempted to bail out. He had great difficulty in bailing out, due to the speed and unmaneuverability of the plane. He had to sit right down in the fire in order to maneuver the plane to a position from which he could bale out. This resulted in severe 2nd and 3rd degree burns to both his legs, from his thigh to his ankles, his right arm and hand, his head and face." Cupp was evacuated and hospitalized for 19 months (he flew combat again in Korea). The Betty he had attacked didn't get far – Lts. Avey and Walley teamed up to shoot it down.

On **23rd September** 24 SBDs and 12 TBFs bombed antiaircraft positions at Jakohina on Bougainville, just west of Kahili. The bombers' escort consisted of 16 P-39s, 16 P-40s and 32 Corsairs of VMF-213 *Hellhawks* and VMF-214 *Black Sheep*. Shortly beforehand B-24s (escorted

F4U-1 Corsairs of VMF-217 on Bougainville in early 1944.

by P-38s) had bombed Kahili and stirred up lots of Zeros. In the ensuing fight Lt. Wilbur Thomas of VMF-213 shot down three Zeros, Lt. Edward Shaw (also a high-scoring ace) got one, and they both jointly knocked down another Zero, for a total of five. On the return leg Thomas had to bail out after the engine of his shot-up Corsair had quit; a few hours later he was picked up by a Dumbo (air-sea rescue plane). The squadron lost Lt. Roberts, who went missing.

The *Black Sheep* returned with four victories and no losses of their own, but there were several close calls. Lt. John Bolt landed with one wing ripped open by a cannon round, but with his first confirmed 'kills' – in Shortland Islands area he had bagged two Zeros. Maj. Stan Bailey saw seven Japanese fighters attempting to machinegun a parachutist (probably a P-40 pilot) in the vicinity of the Shortlands. He dived into them and ran them off a few times but three of them got on his tail and almost downed him, forcing him to dive out and nurse his plane home. He landed with four 20 mm and thirty 7.7 holes in his plane.

On **2nd October** VF(N)-75, the first Corsair night fighter squadron, went into action. Equipped with F4U-2 model and stationed at Munda, this miniature unit, with only six machines on strength, was commanded by LtCdr William 'Gus' Widhelm. The squadron's premier victory came on the last night of the month, when Lt. Hugh O'Neill shot down a Betty bomber.[7]

On **4th October** Maj. Boyington led two divisions of his outfit off from the Russell airstrip. The eight Corsairs were to act as medium cover for a strike by 20 SBDs on Malabeta Hill – a strong antiaircraft position NW of Kahili airdrome. Two Corsairs had to turn back due to mechanical malfunctions.

Arriving 20 minutes early, the six planes flew in lazy circles through the clouds, awaiting the arrival of the bombers. However, no bombers appeared, so Boyington decided to move on in an effort to locate them – and, failing in that, hoping to run across a few Zeros. He was successful in the latter desire – 50 miles south of Kahili they could see clouds of dust rising from the Kahili strip as the Zeros rose to meet them. There were 30 Zeros in all, climbing in flights of 4 and 8. Undeterred, the six *Black Sheep* swooped down, scattering the enemy. Boyington shot down three Zeros – in 60 seconds.

On **11th October**, in the Shortlands area, VMF-213 *Hellhawks* fought their last battle over the Solomons, tallying four Zeros – three by Lt. Wilbur Thomas and one by Lt. Edward Shaw. Thomas, who became a 'triple ace' on that day, reported:

"I was leading the second section. We were weaving over the bombers at 22,000 feet, over the target area [Kahili], when we first made contact with the Zeros. I saw about 15 Zeros. As we scissored, while the bombers were going in on their target and retiring toward their rendezvous area, we made a number of passes at the Zekes, chasing them off each other's tails. I saw Shaw scissor to the right, pass under me and shoot a Zero that had just sneaked in on

F4U-1A of an unidentified USMC squadron taxiing at Torokina airstrip, Bougainville.

my tail. This Zero dove down about 1000 feet, smoking, and then burst into flames.

After shooting down this Zero, Shaw pulled back into formation again. We then saw below us, down close to the water, a P-40 that was being chased by 2 or 3 Zekes. I could see the Jap bullets hitting the water under him. Further to the rear was a lone Zero which I dove down behind, got on its tail, and fired at it at very close range. This Zero made a wingover to the left and crashed into the water between Moila Point on Bougainville and the northern tip of Shortland Island.

I regained about 8,000 feet altitude when I saw two other F4Us on which I joined. They turned out to be Shaw and Handschy. Soon after I had rejoined them I saw, below and behind me, Zeros and P-40s dogfighting, so I turned to lend a hand. I made a high side run on a Zeke that was on the tail of a P-40; it burst into flames and went down, over the southwestern portion of Shortland Island."

Thomas was not over yet. He regained altitude and soon spotted another opportunity: "We were at this time at 9000 feet, with the P-40s below at 5000 feet. The bombers were a good distance ahead of us. Zeros were trailing, out of range. When at a point this side of the Shortland Islands, I saw two Zekes below, trying to sneak in on the tail of the P-40s. I nosed over and went down on the rearmost of these two Zekes. The other was about one quarter of a mile in front of the one I dove on. I fired into this Zeke when at about 5,000 feet, it burst into flames, turned abruptly to the left, and spun down towards the water out of control.

The other Zero started to turn back, apparently having seen the first one go down in flames. I turned inside of it and got on its tail. The Zero turned and rolled in violent evasive maneuvers, but I was able to stay on its tail. On closing on this Zero I gave it three or four shorts and when in close range, I pressed the button for a good long burst, but only one of my guns would fire. Smoke poured from beneath the cowling around the forward area of the fuselage. It dove down through a cloud and I didn't follow it."

Two days later VMF-213, relieved at Munda by VMF-221, completed its third and last tour in the Solomons. The *Hellhawks* left for home with a tally of 104 air victories (76 Zeros, 8 dive bombers, 10 floatplanes and 10 twin-engined bombers). By then eight of the squadron's pilots perished in action and four in accidents; one was captured, and eight were evacuated because of injuries or health problems. The squadron lost 27 Corsairs: 21 in combat and 6 in crashes.

On **15th October** 21 B-24s struck at Kahili. Their escort consisted of 16 Lightnings, 8 Corsairs from VMF-214 and 8 more from VMF-221. Moreover, Maj. Boyington led 4-plane division to act as a fighter sweep to protect tail of returning bombers. The bombers were late and consequently, the fighter sweep arrived over the target before the strike. Orbiting over Kahili, Boyington's division observed 15-20 Japanese fighters scramble. The Corsairs dived down and

This F4U-1A of VMF-212 apparently rolled off the PSP matting and nosed over in the soft ground.

jumped them at 4,000 feet, halfway between Ballale and Kahili. The four *Black Sheep* instantly shot down six Zeros, Lt. William Case and Lt. Warren Emrich two apiece.

So long as the Japanese at Bougainville were willing to come up and fight, the AirSols duly arranged fighter sweeps designed to wear them down. On **17th October** Maj. Boyington led 14 Corsairs of his squadron and 7 of VMF-221 over Kahili to lure the resident Zeros into the air. As they arrived in the vicinity of Bougainville, one division of VMF-221's planes went down at 6,000 feet as a bait to draw the enemy fighters off the strip. The enemy fighters began to take off as the Corsairs circled the area. VMF-214 war diary related:

"Major Boyington took his flight down to 10,000 feet, where they made their first contact with 15-20 Zeros – all Zekes. The reminder of the formation contacted 15 Zekes at 18,000 feet over Ballale. These had apparently circled in from N. Choiseul. At this time another 15 or 20 Zeros came down from 22,000 from west of the Kahili strip. From then on – for 40 minutes – the 21 Corsairs battled 40-50 Zeros all over the sky in an area ranging from Kahili to Ballale, and from Fauro Island to the Shortlands (about 375 sq. miles)". The outnumbered Corsair force won a tremendous victory. The *Black Sheep* bagged 12 Zeros, including three by Boyington; Lts. Magee, Tucker and Heier each got two. VMF-221 contributed with six victories, including two by Capt. William Snider. The Americans recorded no losses of their own.

On **18th October** the target was Ballale airfield. Coordinated in this morning strike were 30 SBDs covered by 24 Airacobras and 12 Corsairs of VMF-214; 12 TBFs covered by 8 New Zealand Kittyhawks and 12 Corsairs of VMF-221; 27 B-24s covered by 16 Lightnings. For some reason the Japanese decided against challenging this raid. The only two victories were scored by Lt. McClurg of the *Black Sheep*.

In the afternoon Boyington, at the head of 12 Corsairs, went back to Kahili, taking along eight Corsairs of VMF-221. The Japanese made a costly mistake of scrambling their fighters when the Americans were already circling over their heads (as the story has it, Boyington taunted them over the radio with verbal abuse until some 20 fighters took off). The Zeros never had a chance to climb higher than 6,000 feet. The *Black Sheep* knocked down eight, including three by Lt. Christopher Magee. VMF-221 contributed with seven, including three by 2/Lt. Jack Pittman and two by Maj. Nathan Post, the squadron's commander (but lost Lt. Schneider, who failed to return). Maj. Post reported:

"When the Zeros took off, they flew straight out over the water and formed up, climbing. At this point we jumped them. We made a pass down through them. I got one from overhead going straight down. As I pulled up I saw I had one on my tail; somebody, I don't know who it was, shot him off. I then checked my guns and found that none were working. I charged them a couple of times and got four working. I then went back into the fight and got another Zero."

VMF-221 achieved this and similar successes despite some serious problems with the aircraft they flew. The squadron's diary noted: "The planes assigned to VMF-221 are definitely not suitable for combat due to excessively hard usage and poor upkeep. The majority need major overhauls. The guns have not been kept up properly. Some of the wing tanks do not draw, so they are impossible to empty. The blowers do not function properly." Lt. Segal, one of the squadron's aces, reported on that day: "I checked my guns when I got up there but couldn't make any of them work. I did not dive but kept on working on my guns. They never did work and I came on home."

A few days later Boyington and his pilots headed for Australia, to rest and recuperate. During their first six weeks in combat the *Black Sheep* had racked up 57 confirmed victories. They lost four pilots killed or missing in action, and two more evacuated because of their injuries.

In late October the resident Japanese air force was conspicuous by its absence. VMF-215, which on **20th October** moved forward to Vella Lavella, had one probable Zeke to show for all its effort by the end of the month. The enemy, however, was far from beaten. In mid-October allied scout planes had discovered nearly 300 combat planes in Rabaul, and another 70 at Bougainville. These were machines of the 11th Air Fleet. Furthermore, in the last days of the month 173 planes (including 82 Zeros), detached from the carriers *Zuiho*, *Shokaku* and *Zuikaku*, arrived in Rabaul. The Japanese were building up their air assets for one more offensive move – operation Ro-Go, very much a repetition of operation I-Go they had launched back in April, which had been a failed attempt to regain air supremacy over the Solomons.

Incidentally, the Allies were also planning a major offensive. On **27th October** New Zealanders seized Treasury Islands, located only some 40 miles from the Japanese base at Kahili. On the same day VF-17 *Jolly Rogers* arrived and flew their first combat mission, a CAP over the Treasuries. On the eve of Bougainville landings the AirSols Corsair force was as follows: VMF-211 at Banika on Russell Islands; VF-17 at Ondonga on New Georgia; VMF-212, -215 and -221 at Barakoma on Vella Lavella – 116 Corsairs in all (36 in VF-17 and 20 in each of the Marine squadrons).

Aircraft of VMF-214 *Black Sheep*, in the foreground F4U-1A (BuNo 17735); Bougainville, February 1944.

At dawn of **1st November** Marines landed at Cape Torokina, in Empress Augusta Bay on the west coast of Bougainville. As was expected, the Japanese vigorously opposed the landings. Shortly after 07:00 hrs the invasion fleet was attacked by 9 Vals escorted by 44 Zeros. At that time eight Corsairs of VF-17 and eight RNZAF Kittyhawks were on station. The *Jolly Rogers* scored their first victories – five Zeros, including two by their commander, LtCdr Blackburn. In the course of another patrol over the bay LtCdr Roger Hedrick, the squadron's executive officer, shot down a Zero. Shortly afterwards VF-17 suffered its first casualty – Lt(jg) Keith was hit by ground fire over Ballale, ditched at sea and was never seen again.

In the early afternoon VMF-215 pilots shot down a Kate and four Zeros in a scrap over Empress Augusta Bay. Lt. Robert Hanson, the future top Corsair ace, got a Kate and two Zeros but ended up in water. Reunited with his squadron a few days later, he reported:

"I saw a flight of 6 Zekes coming out of a trough in the clouds from the direction of Kieta. I also saw about 20-30 planes following this flight. I picked one of this first flight that seemed to be diving for the beach at Augusta Bay. I tailed in behind him in his dive and gave him a burst. He seemed to try to pull away to the left, but I think he had too much speed to maneuver. I gave him another burst and he started smoking and burst in flames. He slowed down and, as I passed over him, pulled his nose up and snapped a few tracers at me but missed. Then he fell off and down, burning as he went.

I banked to the right, then left, and got into position behind another Zeke. When I was sure I had him in my sights, I gave him a long burst. He smoked slightly, then exploded. I pulled up in a chandelle to the left and climbed to 8,000 feet. There I saw about six Kates above and to my left. I had enough speed to make a low-side beam run on the nearest. He peeled off to the right as I shot. I saw no smoke or damage to this one. After I pulled through this run, I ended up above the remaining planes and to their left. I started a fairly low, high-side run on the left plane of the formation. In the middle of my run, they dropped their bombs in the water of Augusta Bay. I finished my run on this left plane and all but the one I shot peeled off to the right. After this first long burst, the Kate nosed over very slightly. Then this dive became steeper and steeper. He did not burn or smoke. I followed him down in his dive, weaving from side to side, picking at him with 2 to 4 guns working spasmodically. I followed him down until he crashed in Augusta Bay, then pulled up and right

Corsairs of VMF-222 getting ready to launch from Vella Lavella, January 1944. Apparently the ground crew member is 'pulling through' the propeller prior to starting the engine in order to clear oil from cylinders.

F4U-1A 'Mary Jo' of VMF-212 parked at Vella Lavella, January 1944.

to chase the other Kates that had peeled off on my first run.

Then I realized that my engine was dead. I had no power and lost speed rapidly. I knew I would have to make a water landing. I believe that on my first pass at the Kates, or when I followed the one down to the water, a rear gunner must have damaged my engine, although I saw no tracers." After a few hours of drifting around the bay Hanson was picked up by destroyer USS Sigourney.

On **2nd November**, another day of fending off air attacks on the invasion fleet, Capt. James Swett of VMF-221 shot down two Vals.

Meanwhile, the Japanese Navy was building up its surface force to strike at the allied invasion fleet in Empress Augusta Bay. To this end Admiral Koga dispatched seven heavy cruisers from Truk. In order to counter the threat posed by this force, the US Navy mounted a preemptive strike against Rabaul anchorage on **5th November**. Two carriers, USS Saratoga and USS Princeton (Task Group 38) launched all of their TBFs, SBDs and Hellcats. While they were performing their mission, VF-17 flew three negative two-hour patrols (24 sorties) over the carriers.

After the first US Navy raid on Rabaul, during which several cruisers had been damaged, the Japanese withdrew some of their warships to Truk. In order to lift this threat once and for all, the Americans staged another, more powerful attack on **11th November**, which involved five carriers from two Task Groups. The second one, Task Group 50.3, was built around three carriers, USS Independence, Bunker Hill and Essex. As before, VF-17 was to cover the ships while their own aircraft were away; 12 Hellcats of VF-33 based at Segi on New Georgia were also involved in the task. Whilst the two covering squadrons were refueling aboard the carriers, Corsairs of VMF-212 and -221 were to relieve them.

While on station over the fleet, LtCdr Blackburn and Ens. Frederick Streig each shot down a Tony snooping around the carriers. At a quarter past 13:00 hrs, when the Task Group 50.3 carriers were still busy taking aboard their aircraft, the ship radars detected an approaching wave of enemy raiders, still some 120 miles distant. By that time the carriers' own Hellcats had been rearmed and refueled, so they were launched back into the air at a frenzied pace; 64 managed to take off. It was a timely reinforcement, for the Corsairs of VF-17 were already running low on fuel. The enemy raid consisted of three groups of aircraft, totaling some 40-50 Vals and Kates escorted by 50-60 Zeros and Tonys. Some Bettys had also been thrown in for good measure.

Before the carrier-borne Hellcats could gain altitude, the *Jolly Rogers* went into action, netting 18.5 victories. Ens. Ira Kepford and Lt. Thaddeus Bell bounced a double string of Vals just as the bombers had begun to tip over and dive towards the ships. Bell got two, Kepford three. Their fuel reserves were dwindling, so they had set out on the long haul back to Ondonga when a second wave of Japanese raiders turned up. This time Kepford shot down a Kate, and almost certainly saved USS Bunker Hill from a torpedo hit. By this time both Bell and Kepford were so low on fuel they had no alternative but to land aboard the carrier as soon as it was clear. Kepford was not the only future VF-17 ace to draw his first blood that day. Lt(jg)

F4U-1A Corsairs of VMF-214 at Bougainville, early 1944.

Howard Burriss bagged a Betty and a Kate, and shared another Kate with a Hellcat pilot. Ens. Streig shot down Tony and Zero apiece. None of VF-17's machines was shot down; two pilots ran out of fuel and had to ditch off the Solomons.

For the following five days Japanese were nowhere to be seen over the Solomons, at least in daylight. At dawn of **17th November** an American convoy off Bougainville was attacked by D4Y Judy dive bombers, which sank one transport. A few hours later, over Cape Torokina, AirSols Corsairs intercepted another such raid comprising 10 Judys and 55 escorting fighters. Two VMF-221 aces, Maj. Nathan Post (the squadron's commander) and his wingman, Lt. Harold Segal, each shot down three Judys. Maj. Post reported:

"They looked like F6Fs and I flew in near them before I saw the meatball. I pulled up and them made a run on them. The first burst I fired burned the first ship, and I swung right over to the second – which also burned brightly at the wing roots and under the fuselage. The last one tried to dive away – I chased him almost to the water before I fired him. This one had a gun mounted on the left side of the canopy which fired back at me. It seemed to be a fixed gun and there did not seem to be a rear gunner."

Lt. Segal reported: "After I saw the Major get his third Jap, I figured he was O.K., so I dove on the three I saw." Since the remaining three Judys doggedly pressed on towards the task force ships, Segal quickly shot them down. He was a little more precise in his description: "The planes seen were small, low mid-wing monoplanes with an in-line engine having a pronounced scoop under the nose, the vertical stabilizer had a squared top. Wing tips were rounded."

It was the last fight of VMF-221 in the Solomons campaign. The following day the squadron left for the USA. During their second and third tour, when they flew Corsairs, VMF-221 tallied 52 air victories for the loss of only two pilots.

During the same morning of **17th November**, in the course of a routine CAP over Empress Augusta Bay, two divisions under LtCdr Hedrick intercepted eight Kates escorted by a four Zeros and five Tonys. Most unfortunately for VF-17, the escorts were led by none other than *Hiko Heisocho* Tetsuzo Iwamoto, a veteran of the air war over China and, according to many sources, the top ranking IJNAF ace by the end of the war. The Zeros skillfully engaged the Corsairs. Lt(jg) Anderson's machine burst into flames and hurtled into the sea. Anderson bailed out at the last possible moment and hit the water at high speed, suffering a few broken bones and an injured spine.

Five minutes after the first skirmish over the Bay a mixed formation of ten Zeros and To nys turned up. Again, the engagement was short and violent. Lt. Clement Gile methodically performed three slashing attacks, each time leaving a flaming Zero in his wake. Lt(jg) Paul Cordray also had a field day, scoring his first confirmed victories – a bomber and two Tonys. Lt(jg) Jack Chasnoff downed two Zeros. Overall, VF-17 came out of this fight with ten victory credits, but at the price of two pilots. Ens. Baker went missing. Anderson was picked up the following day; due to his extensive injuries, he was sent back to the States. It is highly probable that Iwamoto himself, who claimed two Corsairs after this fight, shot down both men.

As time passed, the enemy raiders became increasingly elusive. They were known to carry out sneak dawn attacks on allied shipping in Empress Augusta Bay. On **21st November** two divisions led by Lt. Merl Davenport (VF-17's maintenance officer) were on time to intercept six Zeros armed with bombs – and shot down every one of them. Two were credited to Davenport. The next few days brought no contact with the enemy in the air.

On **2nd December** the *Jolly Rogers* completed their first combat tour and headed for the bars and women of Australia. The Navy could feel proud of its only daytime Corsair outfit. By that time VF-17 had amassed 47.5 victories in the air, destroyed two Bettys on the ground, and sunk two small cargo ships and 11 barges. Its own losses totaled eight Corsairs (two shot down in air combat, three brought down by ground fire, two ditched due to fuel starvation, and one crashed in a flying accident) and three pilots (one shot down by fighters and two from the ground). Interestingly, the first six weeks of fighting did not produce a single ace for VF-17, although several pilots were already credited with four victories.

Aerial Offensive against Rabaul (December 1943 - March 1944)

The ultimate objective of the Allied offensive in the Solomons was Rabaul, the main naval and air base of the Imperial Navy in the southwest Pacific. The Japanese captured it from the Australians in January 1942, and turned it into a stronghold. Surrounded by a ring of airfields, it held a garrison of over 110,000 Imperial Army and Navy soldiers. Rabaul is located at the easternmost end of New Britain, which together with adjacent New Ireland forms Bismarck Archipelago. During the war this natural, semicircular defensive line was better known as the Bismarck Barrier. In August 1943 the Allies decided to bypass and isolate Rabaul rather than attempt to capture it.

There were five airfields around Rabaul, of which four – Lakunai, Vunakanau, Tobera and Rapopo – were used operationally. In mid-December 1943 some 300 aircraft were stationed there. The core of this force was made up of

F4U-1A Corsairs of VMF-214 flying in formation; Bougainville, early 1944.

LtCdr Roger Hedrick, the executive officer of VF-17; Bougainville, February 1944.

three Zero fighter units – 201st, 204th and 253rd Kokutais. Among the Zero pilots who defended Rabaul were several seasoned veterans like the aforementioned Tetsuzo Iwamoto of 253rd Kokutai, who in the course of two months of fighting in Rabaul area claimed 40 victories, including 11 Corsairs. At the turn of 1943/44 experience was practically the only advantage the Rabaul defenders still had. As Henry Sakaida (a renowned historian of the IJNAF) observed,

"The Zero pilots on Rabaul were eventually caught in no-win situation. Their aircraft were ultimately outnumbered and outclassed in performance by the Vought F4U Corsair, Grumman F6F Hellcat and the Lockheed P-38 Lightning. Of these three types of American aircraft, the Japanese pilots considered the Corsair to be the most troublesome, due to its superior speed.

(...) Rabaul started receiving the newest model of the standard Zero, the A6M5 Model 52s in late 1943. The Model 52 had a maximum speed of 351 mph at 19,685 feet. The F4U-1 Corsair's top speed of 417 mph at 19,000 feet literally took the Japanese pilots' breath away".[8]

Rabaul was repeatedly bombed by the Americans from late 1942 onwards, initially by the 5th US Army Air Force stationed in New Guinea. After 11th November 1943 both the 5th AF and the US Navy fast carrier groups moved on, losing interest in Rabaul. Taking over the offensive, the AirSols had some 270 fighters on strength. However, over 100 of them were outdated P-39s and P-40s. Hence, it was the 71 Corsairs and 58 Hellcats that bore the brunt of the fighting. Since there were always some aircraft out of commission, the actual number of Corsairs available for operations at any one time rarely exceeded 50.

In late November VMF-214 *Black Sheep* returned to frontline duties, relieving VMF-212. At Barakoma they shared the airstrip with VMF-223, which had just returned to combat after converting onto Corsairs. For the next two weeks, despite flying daily patrols over Bougainville, the *Black Sheep* failed to encounter any opposition in the air. By that time the IJNAF had given up on the Solomons and had pulled back to Rabaul. Ballale, Kahili and Kara airfields still bristled with Japanese anti-aircraft guns, but aircraft no longer operated from them.

On **10th December** VMF-216 *Bulldogs* moved to an airstrip at Cape Torokina, the first unit to be stationed there permanently. Before long two more airstrips, known as Piva Uncle and Piva Yoke (after the name of a nearby river), were constructed within the perimeter of the small beachhead on Bougainville. They placed Rabaul within range of allied single-engined fighters and bombers, which constituted the majority of the AirSols equipment (the distance to Rabaul was some 250 miles).

The AirSols offensive against Rabaul commenced on **17th December** 1943. The honor of leading the first fighter sweep over Rabaul was offered to Maj. Boyington. The formation consisted of VMF-214, -216, -222 and -223, each represented by two 4-plane divisions, together with 24 Navy Hellcats and 24 RNZAF Kittyhawks. The fighters topped up their fuel tanks at Torokina airstrip, then took a northwesterly course, passing Bougainville and Buka to the right. Ahead, the open ocean stretched out as far as the eye could see.

The New Zealanders elected to press straight on to the target, rather than wait for the American fighters to form up; their Kittyhawks had little fuel to spare. Over Rabaul they were bounced by some 30 Zeros and lost their leader. The Americans appeared on the scene a couple of minutes later, too late to participate. As they circled the area, Lt. Robert McClurg of VMF-214 spotted a Rufe over Simpson Harbor, Rabaul's main anchorage. The temptation proved too strong, and down he went, flaming the Rufe in his first pass. Lt. Donald Moore, another of Boyington's subordinates, also took the initiative into his own hands. He noticed a few Zeros flying in line astern far below, swooped down and splashed two in short order. Besides the three victories credited to VMF-214, there were no other successes by Corsair pilots that day.

On **23rd December** AirSols went back to Rabaul, this time with bombers and plenty of fighters. The strike formation comprised 24 B-24s escorted by Corsairs of VMF-214, -216 and -222 (each with 8 planes), and 24 Hellcats. The sweep formation consisted of 28 Lightnings, 8 Corsairs of VMF-214 and 12 more of VMF-223.

The other formation, scheduled to sweep the area an hour and 15 minutes after the strike

arrived over the area just 15 minutes behind the bombers, because the heavies were half an hour behind the schedule and the fighters slightly early. Nonetheless, this timing proved perfect because it permitted the fighters to engage a great many Zeros as the bombers were leaving the area, and thus divert their attention from the heavies.

F4U-1A Corsair of Lt(jg) Ira "Ike" Kepford, the top-ranking ace of VF-17 *Jolly Rogers* (16 victories), on the ground and in flight. Duct-taped cowl seams are seen to advantage.

Ike Kepford himself.

Boyington was in his element. He shot down four Zeros: three over St. George's Channel and the last one over Simpson Harbor. Three of his subordinates, Lts. John Bolt, William Heier and Robert McClurg, scored double victories. In the section titled „Comparative performance, own and enemy aircraft" of the daily report, Frank Walton, the squadron's intelligence officer, remarked: "Nothing new, Zekes still burn". However, it was a costly day for the *Black Sheep* – the price for the squadron's 12 victories was three pilots MIA.

The remaining three Corsair squadrons returned safely to base with no one missing. Maj. Marion Carl, the CO of VMF-223, shot down a Tony for his first Corsair victory (he already had 16.5 victories to his credit, scored on Wildcats in 1942). He recalled:

"On 23 December I led my guys up to Torokina again, where we topped off and launched for Rabaul as I led twenty Corsairs and twenty-eight P-38s. We made contact over Cape St. George and I stalked a new opponent through the clouds. It was a Kawasaki Tony, a sleek, good-looking Japanese Army fighter that resembled the Messerschmitt 109. I splashed him between Rabaul and New Ireland while the rest of the squadron claimed three more confirmed and three probable without loss. Boyington's squadron and VMF-222 added fifteen more; we were taking a toll on Rabaul's defenders".[9]

Another high-scoring ace of VMF-223, Capt. Kenneth Frazier, scored his first and only Corsair victory, shooting down a Hamp over east coast of New Ireland. In all, Carl's squadron got four victories that day – their first on Corsairs.[10]

On **25th December** US Navy carriers struck at Kavieng. AirSols was tasked with tying down Rabaul-based fighters. A formation of 24 B-24s, escorted by 64 Lightnings, Corsairs, Hellcats and Kittyhawks (16 of each type), did the job. Approximately 30 Japanese fighters intercepted them over Blanche Bay and the Duke of York Islands. In the shootout that followed, VMF-214 got four victories, and VMF-223 three (including two Zekes by Capt. Harlan Stewart), but lost a pilot, Lt. Sahl, believed shot down at sea.

On **27th December** 39 Corsairs (of VMF-214, -216, -223 and -321) and 20 Hellcats went over to Rabaul to goad the resident fighters into the air. Corsair pilots racked up 15 victories, of which 7 were credited to VMF-216. The *Bulldogs* owned much of their first success to Boyington, who led them into a perfect position for a 'bounce'. As VMF-216 diary relates: "When planes in sweep were about 5 miles south east of Rabaul, enemy planes were observed taking off in two plane sections from Lakunai Aerodrome. Major Boyington, tactical flight commander in charge of sweep, radioed the following instructions: 'Take it easy and let down slowly. The Japs should be up here in about four minutes'. Our planes were at 21,000 ft. They continued on over target in a wide 360° turn to the left, slowly dropping to about 18,000 ft. by the time they completed turn. This brought them out about 1,000 ft. above and behind 10 or 12 Zekes." From then on each pilot knew himself what to do! The *Black Sheep* had a successful day, too – they bagged six victories, including two by Lt. Don Fisher.

On **28th December** AirSols fighters again turned up over Rabaul, with yet another sweep. There were 44 machines from VMF-214, -216, -223 and -321, led by Maj. Rivers Morrell, the commander of VMF-216 *Bulldogs*. It was a sensationally successful day for Morell's squadron – 17 victories (including 'doubles' by Lt. Robert Anderson, Lt. Charles Schwartz and Lt. Roland Marker) for no losses of their own! Schwartz reported his second victory in the following manner:

"About 15 miles from Cape St. George, I was jumped by two Tonys, and three Zekes, from above and behind. For approx. ten minutes they chased me. Had me bracketed. Their fire went in and around my plane. Finally after a futile attempt they turned to the right and headed home. As the last Jap plane turned and was heading home, I turned also, at 6,000'. Infuriated by the damage done to my plane, I pulled up and dead astern on a lone Tony. When I was in range I opened fire, the Tony smoked, swerved off to the right and went into the sea. Immediately I turned about and headed home. (…) My plane was shot up considerably. My left wing, horizontal tail section, and fuselage back of the cockpit had 20 mm holes".

As it happened, all the 24 victories credited to the *Bulldogs* during their first tour were scored on 27th and 28th December. A few days later they left for R&R, and by the time they came back for their second tour, the air battle for Rabaul was practically over.

On the same day Lt. Robert See, the future (and only) ace of VMF-321, flew his first combat mission and scored his premier victory:

"That melee was the wildest of all dogfights. I didn't get a single good shot until I spotted a lone Zero pulling tight chandelle to evade a pursuing Corsair. He was going slow and a set-up, so I dove on him. When I fired, I expected six 50-calibers to blaze out, but only one gun fired intermittently. Now I could visualize an unhappy ending to this otherwise perfect bounce, with the Zero turning on my tail and gunning me. Despite this potential threat, I kept boring in on him and suddenly the left wing of the Zero ignited into a ball of fire".[11]

F4U-1A Corsair of VMF-212 at an unknown location, 1944.

Kepford taxies out of revetment for strike on Rabaul; 19th February 1944.

VMF-223 contributed with four victories, including three Zekes by Capt. Fred Gutt. It was a bad day for the *Black Sheep*, who scored four victories, but lost three pilots.

On **30th December** the weather was so bad that rendezvous proved impossible, and the fighters – only 11 Corsairs and 12 Hellcats – just pressed on in the general direction of Rabaul. Perhaps fortunately, the opposition in the air was meager. Capt. Edwin Olander of VMF-214 shot down a Zero in a brief skirmish over St. George's Channel; it was his fifth and last victory.

The new 1944 year began very much like 1943 ended. On **2nd January** VMF-211, which hitherto hadn't seen much action, finally began to score steadily, amassing 71 confirmed air victories by the month's end. The first victory was credited to Maj. Thomas Murto. On that day three 4-plane divisions were off on an early morning fighter sweep over Rabaul. The squadron's diary related:

"While over St. George's Channel at 20,000 feet, Major Murto saw a lone Zeke headed toward Rabaul about 1500 feet below. He nosed over and made a run on the Zeke firing his tracers into the fuselage. The pilot started to get out of his cockpit as though preparing to bail out, but fell back into his cockpit. Major Murto believes he killed the pilot for the Zeke then nosed over and dove down into the bay."

Later that day Maj. Boyington led 36 Corsairs and 20 Hellcats on a fighter sweep over Rabaul. VMF-321 scored five victories, including two Zekes shot down over Simpson Harbor by the squadron's commander, Maj. Edmund Overend (who, like Boyington, was a former 'Flying Tiger'). However, Capt. Blount failed to return from this mission and remains missing to this day.

On **3rd January** Maj. Boyington led a force of 28 Corsairs (of VMF-211, -214 and -223) and 16 Hellcats to tackle Rabaul defenders one more time and hopefully beat the service record of 26 individual victories – something everyone expected of him. Over Rabaul, a solid bank of clouds cast a deep shadow on the ground, and ground haze further reduced visibility. Meanwhile, 70 Zeros of 204th and 253rd Kokutais had scrambled from Rabaul.

When Boyington spotted six Zeros below, he eagerly rolled over and swooped down. His wingman, Capt. George Ashmun, faithfully followed him. Boyington's first victory of the day was witnessed and confirmed by other pilots. Then something went terribly wrong. The rest of the pilots either lost sight of the two diving Corsairs in the murk below, or were tied up by other Zeros. Oblivious to the mounting danger, Boyington pressed on. In the area of Rapopo airfield the two Corsairs were cornered by some 20 Zeros. Boyington later claimed that he shot down two more Zeros, before he and his wingman were overwhelmed and gunned down. Ashmun was killed; Boyington remained in captivity until VJ-Day. During that time he was 'posthumously' awarded the Medal of Honor and promoted to the rank of Colonel. The squadron was taken over by his executive officer, Maj. Henry Miller, but shortly afterwards the *Black Sheep* were disbanded (during their two six-week combat tours in the Solomons they amassed 97 confirmed victories).

Altogether, the day's fighting produced nine destroyed credits for the AirSols. Two of them were credited to VMF-211. Capt. Robert Hopkins, cruising at 27,000 feet over Blanche Harbor, saw three Zekes at 24,000 feet flying opposite course. Squadron's diary: "He made a high side run on number two plane, and his wingman, 2nd Lt. R.D. Rushlow, made a run on the third plane. The second plane apparently did not see Capt. Hopkins for he continued on course. His tracers entered the fuselage in the cockpit area and he saw the pilot attempt to crawl out when his canopy had been shot off. The Zeke then nosed over and dove down with the pilot half hanging out of the cockpit, apparently dead." A few days later Hopkins was promoted to a Major; he was killed in a mid-air collision before the month was out.

On **4th January** AirSols fighters – Marine Corsairs (VMF-211, -214, -223 and -321) and

Navy Hellcats – shot down 10 fighters over Rabaul. Lt. George Dixon of VMF-321 got two. He did it by attacking by himself three Zekes flying low in tight formation over Cape St. George, and flaming two in a long, spraying burst. The squadron lost Capt. Carter, who went missing over Simpson Harbor and his ultimate fate is unknown.

On **5th January** 84 AirSols fighters, most of them Corsairs, provided cover for SBD and TBF strike on Rabaul shipping. Because of poor weather, all planes of the strike were forced to turn back after having gone as far as the eastern coast of New Ireland. The following day flight echelon of VMF-215, back from R&R, reported to Barakoma strip on Vella Lavella, relieving VMF-223.

On **7th January** VMF-211, -215 and -321 were among AirSols fighters shepherding TBFs and SBDs on a strike mission to Rabaul. VMF-211 scored three victories, as did VMF-321. VMF-215 got one – with a bit of luck. Capt. Richard Braun, one of the squadron's aces, reported: "Three Zekes got on my tail and I dove through the cloud cover to get away from them. When I came through the cloud cover, I was right on the tail of a Zeke, which I set aflame with one short burst."

On **8th January** a four-plane division of VMF-211, led by Maj. William Campbell, was sent on an early morning search mission in the Buka area. At Ramun Bay they sighted three Japanese float planes and two barges with many personnel aboard. All the planes and barges were thoroughly strafed, set on fire and sunk. By that time, however, Major Campbell was nowhere in sight and he remained missing in action ever since.

Form ACA-1
Sheet 4 of 5

AIRCRAFT ACTION REPORT

RESTRICTED
(Reclassify when filled out)

XII. TACTICAL AND OPERATIONAL DATA. (Narrative and comment. Describe action fully and comment freely, following applicable items in check list at left. Use additional sheets if necessary.)

ENGAGEMENT WITH ENEMY
OWN AIRCRAFT
Disposition
Altitudes
Speeds
Approach Tactics
Use of Cover, Deception
Angles of Attack and Their Effectiveness
Distance of Opening Fire
Defense Tactics and Their Effectiveness
ENEMY AIRCRAFT
Method of Locating, Distance
Disposition
Altitudes
Speeds
Approach Tactics
Use of Cover, Deception
Angles of Attack
Distance of Opening Fire
Defensive Tactics
COMMENTS AND RECOMMENDATIONS
Own Weaknesses
Enemy Weaknesses
Offensive Tactics, Own
" " , Enemy
Defensive Tactics, Own
" " , Enemy
Flexible Gunnery, Own
Escort Tactics
Fighter Direction
Use of Radar
Night Fighting
Recognition, Aircraft
ATTACK
OWN TACTICS
Method of Locating Target
Approach to Target
Altitudes, Speeds
Approach
Dive
Release
Pull-Out
Dive Angle
Strafing
Retirement
Defensive Tactics
DEFENSE, ENEMY
Evasive Tactics, Ships
Concealment
Anti-aircraft
Searchlights
Night Fighter Tactics
COMMENTS AND RECOMMENDATIONS
Bombing Tactics
Torpedo Tactics
Effectiveness of Bombs, Torpedoes
Selection of Targets
Fuzing
Strafing Tactics
Defensive Tactics
Use of Radar
Reconnaissance
Photography
Briefing
OPERATIONAL
Navigation
Homing
Rendezvous
Recognition, Ships
Communications
Flight Operations
Search and Tracking
Base Operations
Maintenance

The formation of approximately 50 fighter planes with Major Boyington of VMF 214 as tactical commander, headed up the W. coast of Bougainville, arriving over Cape St. George stacked at 20, 22, and 24 thousand feet.
Boyington led the formation in a huge right turn around Rabaul at 0800. After a 180° turn, contact was made by Boyingtons division with a formation of 12 Zekes at 19,000 feet, off Rapopo at 0815.
At this time Major Boyington was leading a division made up of Capt. Ashmun his wing man; Lt. Matheson leading his second section, with Lt. Chatham as Mathesons wing man. These 4 had joined when 2 men out of each of the original 2 divisions had been forced to return to base with various types of motor trouble.
The 2 sections of Boyington's division split up as they initiated their runs on the formation of 12 Zekes climbing toward them. Boyington went down in a high six olcock run with Ashmun protecting his tail, while Matheson and Chatham went on a pair of stragglers. Chatham's guns failed to fire as he settled into position so he climbed up and away and saw both Boyington's and Matheson's Zekes go down burning.
Matheson nabbed his in a high headon run, opening fire at 200 yards and knocking pieces off the cowl and motor of the enemy plane. Matheson then pulled up hard to the right and watched his zero go down burning, At the same time he saw Boyington's Zeke going in.
VMF 223 pilots confirm both planes.
Matheson circled for same 20 minutes looking for a Corsair which to join but was unable to locate one because of the haze which made visibility poor. His wingman had returned to base after the first pass when his guns failed to work due to an electrical failure.
No further sight or radio contact was made with either Boyington or Ashmun after this initial pass - although someone called Dane base and reported that he was going to have to make a water landing.
This was Major Boyington's 26th enemy plane - every one a fighter plane - and every one destroyed over enemy territory. It ties the record established in the last war by Capt. Eddie Richenbacker and tied at Guadalcanal by Marine Corps Major Joe Foss.
10 large and one especially large ships (transports or Cargo ships) were observed in Simpson Harbor. The largest one threw up a great deal of A.A.
A.A. otherwise was intense, heavy caliber and quite accurate.

CONFIDENTIAL

7

ALLBET - MFD BY THE EGRY REGISTER CO., PATENTED

The *Black Sheep* lose their commander – action report from VMF-214 war diary of Maj. Boyington's last mission, 3rd January 1944.

LtCdr Blackburn, the commanding officer of VF-17 *Jolly Rogers*, by the squadron's scoreboard; Piva Yoke airstrip, Bougainville, February 1944.

On **9th January** AirSols attacked Tobera, one of Rabaul's airfields. The strike force consisted of 24 SBDs and 15 TBFs, 34 Corsairs, 16 Hellcats and 20 P-40s. The enemy's force was estimated at 50 fighters. VMF-211 got four. VMF-212 scored eight, including double victories by Lt. Philip Delong and Maj. Hugh Elwood (the squadron's CO). After this engagement the squadron's intelligence officer noted, under the heading 'comparative performance, own and enemy aircraft':

– F4U-1 can pull away from Zeke and Hamp [the A6M3 Type Zero Model 32 with squared off wingtips] in level flight.

– F4U-1 can climb away in steady climb, but Zeke and Hamp can outzoom.

– F4U-1 cannot turn with Zekes & Hamps at high or medium speed.

– F4U-1 can dive away from Zeke, Hamp, and Tony.

– F4U-1 can take much greater damage and return safely than Zeke or Hamp.

– Six fifties of F4U-1's are hard to beat. Neither Zeke or Hamp can bring similar firepower to bear.

On **12th January** 44 Corsairs, 8 Hellcats and 4 Lightnings covered B-24s making for Lakunai. Capt. Donald Aldrich of VMF-215, flying high cover at about 28 to 30,000 feet, easily scored two victories: "I saw 3 Zekes approaching from the west at about 25,000'. We were executing a violent two-plane weave and I swung wide and dove on the Zeros, approaching from astern. My first burst (a fairly long one) started flames along the wing root of the first Zero. I looked back and saw that my wingman [Lt. Burke] had flamed the second one. The third one rolled and dove out.

(...) Very shortly thereafter I saw a pair of strays (Zeros which seemed to have no plan of action but were flying a course paralleling to the bombers). I dove on them from dead astern (they were at about 25,000') and saw one of them go into flames on my first burst".

Occasionally, bad weather hampered operations. On **13th January** SBDs were sent out to strike at Rabaul, but due to the failure of the bombers and fighters to effect a rendezvous, the strike was ordered to return to base. On the same date Lt. Reuben Johns of VF(N)-75 (which in the interim had moved up from Munda to Bougainville to protect the beachhead from night intruders), scored the seventh and last victory credited to the squadron, shooting down a Val over Cape Torokina.

On **14th January** the air battle over Rabaul suddenly flared up. That day VMF-215 pilots were briefed for TBF cover on an SBD-TBF strike to Lakunai. The squadron sortied 22 Corsairs, with 8 RNZAF Kittyhawks in assistance. Since Lakunai was clouded over, the bombers chose instead to hit the shipping in Simpson Harbor and Blanche Bay. The Japanese scrambled 50-70 fighters to oppose this raid, and there was a running fight all the way in and out of the target area. VMF-215 had the best day in the squadron's history – 19 victories. Lt. Robert Hanson got five, Capt. Arthur Warner four, Maj. Owens (the CO), Capt. Robinson and Lt. Cox each bagged two.

Hanson went into the dive with the TBFs and pulled out at 2,000 feet. After he had flamed his first Zero of the day, he and his wingman got separated. Alone and uncomfortably low, Hanson took cover in clouds, popping out from time to time to see if there's anything around to shoot at. With so many Zeros milling around, he succeeded in bagging four more.

On the debit side, three of the squadron's pilots were lost (two MIA, one killed in a crash on takeoff); another was hospitalized for shrapnel wounds and shock after he had bailed at Torokina from his shot-up Corsair.

VMF-211, which flew escort to SBDs, contributed with three victories.

On **15th January** three divisions of VMF-211 escorted a B-25 strike to the Rabaul area. Because of the bad weather the flight was returned to base after having gone well up the northern coast of New Ireland in search of an opening in the front.

In the evening, while flying a local patrol, three of the squadron's pilots intercepted a lone Zero. As the unit's diary bluntly states, "They closed in and then bracketed the Zeke, all three making no deflection shots at will. The Zeke went down to 1,500 feet with both wings on fire, and then the pilot parachuted out. Even tho the pilot appeared to be dead when he hit the water he was strafed to insure the certainty of the kill."

On **17 January** AirSols struck at enemy shipping in Simpson Harbor. The escort to 48 SBDs and 18 TBFs comprised 46 Corsairs (of VMF

211, -212 and -321), 7 Hellcats and 24 Lightnings. The escorting fighters won 18 victories, including 10 by Corsair pilots, but own losses were considerable: five P-38s and F4U, F6F, SBD and TBF apiece.

VMF-212 diary reported: "No interception was made during the attack but the Japs initiated a hard-driving attack when the withdrawing bombers reached Karavia Bay which continued well into St. George's Channel. An estimated 30 Zekes, 4 Hamps and 2 Tonys participated." The report also includes this curious remark: "Two cases of seeing all-Jap dogfights were reported. But none of our pilots accepted the invitation."

VMF-321 diary noted: "Major Overend believes that the F4U is a poor plane for low and close cover. It is not maneuverable enough. The P-40 and F6F can do much better a job."

On **18th January** Tobera airfield was the target for 12 B-25 Mitchells escorted by 70 AirSols fighters: Corsairs, Hellcats and New Zealand's Kittyhawks. VMF-215 tallied 6 Zekes, 2 Hamps and 2 Tonys; Capt. Harold Spears got three, and Lt. Creighton Chandler two.

Non-combat hazards were always present. As VMF-211 diary states on that day, "Taking off on the B-25 escort, Lt. W.R. Culler cracked up in the middle of the runway. The plane was totally damaged, but Lt. Culler received only moderate injuries due perhaps to the fact that he was thrown clear of the plane"(!).

On **20th January** 36 Mitchells struck off for Vunakanau airfield, with 36 Corsairs, 20 Kittyhawks, 8 Lightnings and 8 Hellcats to protect them. The defending force was estimated at 35-40 fighters. VMF-215 tallied six (Zekes and

Form ACA-1
Sheet 4 of 5

AIRCRAFT ACTION REPORT

RESTRICTED
(Reclassify when filled out)

XII. TACTICAL AND OPERATIONAL DATA. (Narrative and comment. Describe action fully and comment freely, following applicable items in check list at left. Use additional sheets if necessary.)

ENGAGEMENT WITH ENEMY — **OWN AIRCRAFT**: Disposition, Altitudes, Speeds, Approach Tactics, Use of Cover, Deception, Angles of Attack and Their Effectiveness, Distance of Opening Fire, Defense Tactics and Their Effectiveness. **ENEMY AIRCRAFT**: Method of Locating, Distance, Disposition, Altitudes, Speeds, Approach Tactics, Use of Cover, Deception, Angles of Attack, Distance of Opening Fire, Defensive Tactics. **COMMENTS AND RECOMMENDATIONS**: Own Weaknesses, Enemy Weaknesses, Offensive Tactics, Own / Enemy, Defensive Tactics, Own / Enemy, Flexible Gunnery, Own, Escort Tactics, Fighter Direction, Use of Radar, Night Fighting, Recognition, Aircraft.

ATTACK — **OWN TACTICS**: Method of Locating Target, Approach to Target, Altitudes, Speeds, Approach, Dive, Release, Pull-Out, Dive Angle, Strafing, Retirement, Defensive Tactics. **DEFENSE, ENEMY**: Evasive Tactics, Ships, Concealment, Anti-aircraft, Searchlights, Night Fighter Tactics. **COMMENTS AND RECOMMENDATIONS**: Bombing Tactics, Torpedo Tactics, Effectiveness of Bombs, Torpedoes, Selection of Targets, Fuzing, Strafing Tactics, Defensive Tactics, Use of Radar, Reconnaissance, Photography, Briefing.

OPERATIONAL — Navigation, Homing, Rendezvous, Recognition, Ships, Communications, Flight Operations, Search and Tracking, Base Operations, Maintenance.

24 F4U-1s of VMF 215 took off from Barakoma at 1900/13 GCT Jan. 13,1944 and landed at Torokina 2010 GCT. The pilots were briefed for TBF cover on an SBD-TBF strike to Lakunai. Took off at 2310 (Capt. Stidger crashed & burned on takeoff; Lt. Fitgerald's ship had been hit by another ship while parked, so he was not able to take off) and all fighters rendezvoused and flew in formation to the bomber rendezvous 10 miles W. of Taiof Island at 6000'. Rendezvoused and on course to the Rabaul area at 0025 GCT.

Major Owens & Capt Braun's divisions flew high cover at 18,000'; Capt. Warner's and Conant's divisions flew medium cover at 16,000'; and Capt. Aldrich's & Langstaff's division flew low cover on the TBFs at 14 to 15,000', with 8 P40s as close cover with the TBFs at 13,000'. This whole flight flew to the rear and slightly to the right of and on same altitude with the SBDs and their cover.

Directly after crossing the coastline of New Ireland (just North of East Cape) a flight of six to eight Jap fighters was spotted at about 9 o'clock at 20,000 feet. These Zekes worked around the formation until they were at 6 o'clock and were constantly being reinforced by others until there were from 50 to 70 above and behind our flight. They were looping, slow-rolling and making an occasional pass at our flight, continuing the fight and pressing harder and harder as it approached the target.

The bombers, due to the fact that Lakunai was largely under a cover of cloud, chose instead to hit the shipping in Simpson Harbor and Blanche Bay. As the bombers started their runs, the whole formation dropped in altitude, keeping a proportionate altitude differential. Hits or near misses were seen on 6 to 8 large ships in Simpson HARBOR. One 2000-lb bomb was seen to hit the stern of a large AK in Blanche Bay just to the North of Raluana Pt. This ship seemed to stand on end and sank almost immediately. Two near misses and one hit were seen on an AK just SE of Praed Point.

The flight retired over the Rapopo area to a rally point 10 miles East of Cape Gazelle. During the retirement, because of the masses of Jap fighters stacked from 3000' to 6000' to the rear of the flight, our planes were forced to telescope into a more compact formation. This was the period of bitterest fighting.

Our planes accounted for 19 Zekes destroyed and 6 probably destroyed, compared to our losses of two Corsairs which did not return from the flight and one Corsair shot up too badly to attempt a landing, from which the pilot parachuted to safety at Torokina.

SECRET

38

VMF-215's best day – aircraft action report and individual reports by Lt. Hanson (5 victories) and Capt. Warner (4 victories), 14th January 1944.

Lt. Hanson: "I went into the dive with the TBFs and pulled out at 2000'and checked my guns. These TBFs seemed to attack shipping in Simpson Harbor, Keravia Bay and Blanche Bay amid very bad heavy A.A. I went to the rally point with TBFs; there were just a few there. I saw some SBDs and an F4U under attack by three Zekes. I ducked into a cloud and came out toward the SBDs - toward the New Ireland side of the rally point area. There were lots of Zekes cruising around low, and their speed did not seem to exceed 180 knots. My wingman, Bowman, was with me at this time; we came up astern of two Zekes at 1,500' and each fired at one of the Zekes. Mine flamed brightly immediately, Bowman's flamed a little and turned into a cloud. I went around the cloud and saw a flamer come out and start down. This must have been Bowman's, for I saw it coming out on just about the course that his had gone in. At this point I lost contact with Bowman. Then I came up on two Zekes which were in front of me and on a parallel course. When they saw me they split and one of them went into a cloud. The cloud wasn't very thick: you could see about 200 yards in it. I followed the one that went into the cloud and chased him down. We came out of the cloud at about 300' and opened fire. I saw my tracers going right into him and before I pulled away, I saw him explode.

"I then pulled up and went back into the cloud and stayed near the edge of it so I could get back in if I had to. I wanted to get all the altitude I could in case I had to cross the strip of clear sky to the rally point, as there were still Øs all around - mostly from 3000' to 6000', but a few of them lower. There must have been 30 of them. At about 2000' I looked out of the cloud again and saw two F4Us at about my altitude coming toward me, with two Øs coming down on them from about 3000'. I put my nose up towards the Øs and they pulled away. The Us must have gone into the cloud for when I came back to join up on them I could not find them.

"The clouds got higher and higher as they got over towards New Ireland, and I stayed near them, playing hide and seek in and out of the clouds - all the time getting altitude. At about 3000' I saw two Øs at 9 o'clock at about 2500' crossing my course. I dove and they turned away from me back towards the clouds. I opened up on one when I had a 45° deflection shot and saw him burst immediately into flame. I pulled out well below them and turned back into the cloud.

"Then I started again to climb, going in and out of the cloud. At 2500' I looked out of the clouds again and saw I was right on the tail of a Ø just a little below him. I ran right up to him and fired. My tracers went right into his belly and he burst immediately into flame. I think that this is the best way to shoot them: from astern and below - the belly seems to be the most vulnerable point.

"I made a few more passes without any results and looking over towards the rally point, I saw none of our planes - there were still a few Zekes flying around apparently without any plan. I saw one over towards where the rally point had been - at about 2 ,000'.

SECRET 39

Hamps), including 'doubles' by Capt. Aldrich and Capt. Spears. Lt. Chandler dived to treetops level to get his third confirmed victory – he shot down a Zeke with its wheels down, as it was about to land at Tobera. VMF-211 tallied six, including two by Lt. John Hundley. VMF-321 contributed with three (two by Capt. Robert See), but lost three pilots. Lt. Marshall was hit by AA 20 mm fire and ditched in St. George's Channel; his plane broke in two upon impact and rapidly sank. Also Capt. McCown and Lt. Brindos were unaccounted for by the day's end.

On **22nd January** 18 Mitchells attacking Lakunai airfield enjoyed the escort of over 90 AirSols fighters, including 27 Corsairs of VMF-215 (being stationed at Barakoma, as usually they refueled at Bougainville beachhead before proceeding to Rabaul), 12 of VMF-211 and 16 of VMF-321. In the ensuing fight Corsair pilots shot downed and killed *Itto Hiko Hei* Hiroshi Shibagaki, one of 204th Kokutai's aces, who had scored all of his 13 victories during two months of combat over Rabaul. VMF-215 pilots tallied 10 victories (three by Lt. Hanson). In VMF-211 the box score for the day was five (again two by Lt. Hundley). The worst off was VMF-321, which scored one victory but lost two pilots (Lts. Wardle and Smith).

On **23rd January** 74 AirSols fighters (including 48 Corsairs) supported 66 SBDs and TBFs in a strike on Lakunai. Enemy fighter force was estimated at 45 Zekes, 12 Hamps and 6 Tonys. VMF-211, -212 and -321 pilots racked up a total of 29 victories. In the afternoon the same three Corsair squadrons performed a sweep over Rabaul, this time bagging 16 – for a total of 45! Notably, in the afternoon the opposition was even heavier – against 32 Corsairs, the Japanese scrambled some 70 fighters.

I dove on him from about 3,000' got a 25° deflection from astern and above him. He started to smoke right away at his wing roots then I saw tongues of flames from his wing roots and engine.

"By that time I was sure that our planes had gone home, so I went back into the clouds and out over New Ireland. Then I saw 2 Øs on my tail. I ducked into the clouds and did a violent right turn and beat it for home. I got to Torokina about 20 minutes after the rest of the planes with 20 gallons of gas and only 400 rounds of ammunition left."

Ø low altitude defence tactic noted: "If a pair of Øs is attacked at low altitude by a single plane from astern, one Ø will pull out to the side and up, the other will pull out and down. In this way the Øs are in a position to be of immediate mutual protection."

Capt. Warner: "After the TBFs started into their dive, the cover followed them in and accompanied them at low altitude to the rally point. There were from 50 to 70 Øs stacked up above and to the rear of the flight that were making repeated attempts to get through to the bombers. I had gotten separated from my wingman out over New Ireland and was covering the rear of the flight in this retirement. I had my controls set at 2450 RPM and 45" and was holding all the speed I could muster thinking to from being attacked by my speed and the violence of my maneuvers.

Just after we crossed Rapopo, a Ø made a shallow highside attack on the TBFs. I dove to intercept him in his recovery and met him almost head-on at about 400' altitude. I got a good burst into his engine and he was flaming badly by the time he went by me. In my recovery I pulled up to 1600' right under a Ø that was flying to the north over the formation. When I first saw him, it was about 90° deflection shot, but I kicked my nose over to follow his course and gave him a long burst when I had a 45° deflection. I saw this one smoke and start down. At 2000' I did a wingover to the right and, looking down, saw the smoking Ø hit the water.

I continued my right turn and crossed over the formation. Then I saw a Ø starting to dive toward the formation from about 3000' - I stuck my nose up at him, but he pulled up and away before I got him in range Then I did a wingover to the left and started across the formation, when I saw another Ø diving in to attack the bombers. I dove right down on top of him from dead astern and gave him a long burst at 1200' altitude. I saw this one hit the water just east of Cape Gazelle.

I stayed down with the bomber cover until we got to the rally point. When the bombers started to circle to form up in the rally point I saw a Ø making a flat run on the tail of an F4U that was on an opposite course to me and about 200' higher. I pulled up into the Ø and got a 90° deflection shot into his belly. He flamed immediately, but I did not see him hit the water. I then went back down to the medium cover and started weaving with them over the bombers on the course home."

40

Honors that day went to VMF-212, which raked up 21 victories, but lost Maj. Donald Boyle, who went missing (he was the squadron's only loss on that day). VMF-211 also scored heavily, tallying 16 victories (including three by Maj. Julius Ireland) for no losses of their own. VMF-321 chipped in six victories, including four by Capt. John Norman; Lt. Marsh failed to return.

On **24th January**, another day of AirSols attacks on shipping in Simpson Harbor, 18 TBFs arrived at the target area well protected by 84 fighters, including Corsairs of VMF-211, -215 and -321. VMF-215, back in the heat of action after one-day break, scored six – including four Zekes by Lt. Hanson, who again stalked his unsuspecting victims under cover of clouds. Maj. Robert Owens (the squadron's commander), coming back in a badly shot-up plane, ditched at sea near Torokina and was rescued by a Dumbo. He was credited with one victory (his seventh) – a Ki-44, by his own account: "He was not like any other Jap fighter plane that I've seen before. It had a rather short stubby fuselage with elliptical wings fairing back into the fuselage; large radial engine with large spinner – and seemed to be going exceptionally fast" (Curiously, according to all known documents Ki-44 Tojos were never stationed in Rabaul).

Due to rough conditions at the forward bases and the hectic pace of operations, serviceability was a mounting problem. In VMF-215, of 19 planes that took off, 6 had to turn back with various malfunctions.

Lt. Harold Segal, who had joined VMF-211 for his third combat tour, shot down two Zeros over Cape Gazelle; these were his last victories, eleventh and twelfth. It was a successful day for VMF-321, which scored seven victories

F4U-1A of VMF-212 photographed at Bougainville in early 1944. Note a mechanic dozing off in shade under the aircraft.

with no losses to themselves (double victories were credited to Lts. Robert Baker and John Buzzard).

The Japanese were not giving up, whatever the cost. On **25th January** Admiral Koga stripped carriers *Junyo*, *Hiyo* and *Ryuho* of their air groups, and dispatched 132 aircraft (including 70 Zeros) to Rabaul.

On that day AirSols Corsairs flew a fighter sweep over Rabaul but apparently the defenders learned to ignore raids without bombers. VMF-211 diary states: "Because of the refusal of the Japs to send up any fighters, no sightings or contacts were made."

On **26th January** AirSols mounted yet another massive airstrike against Lakunai. The formation counted 48 SBDs, 18 TBFs, 22 Corsairs of VMF-215, 32 Corsairs of VF-17, 12 Kittyhawks, 12 Lightnings and 8 Hellcats – 66 bombers and 86 fighters in all.

VMF-215 scored 14 victories (all of them Zekes). Lt. Robert Hanson downed three; double victories were credited to Capts. Harold Spears, Arthur Warner, Donald Aldrich and Edwin Hernan. The squadron's diary proudly states: „All the TBF's we were escorting returned safely and all our planes returned to base at Barakoma by 1730." As usually, there were some close calls. Capt. Warner: "I got hit in the accessory section by 20 or 40 mm and oil gushed back all over me. I put my nose down an headed for home"; Capt. Aldrich: "I got two Zeros on my tail and tried violent evasive tactics. I finally dove 8,000' to 10,000' and lost them – but I almost didn't pull out of my dive, as my left elevator trim tab and my right aileron were pretty badly shot up."

For VF-17 *Jolly Rogers*, fresh from R&R in Australia, it was their first mission over Rabaul. LtCdr Blackburn shot down his fifth enemy aircraft, which made him the squadron's first ace. Two victories were scored by Blackburn's wingman, Lt(jg) Douglas Gutenkunst, who at one moment broke off to bag a stray Zero, then shot down another in self-defense, and in no time returned to guard the commander's tail. Altogether, the *Jolly Rogers* got eight, but they lost two pilots, and a third one was hospitalized for a week with head injuries, after writing off his Corsair in a crash landing.

On the same day VMF-211 was close to losing its top-ranking ace, Lt. Harold Segal. His plane developed an oil leak, which caused the engine to freeze. Segal ditched five miles off the Torokina strip and was quickly picked up. Earlier that month (on the 13th) the squadron had lost Capt. A.R. Vetter in very similar circumstances. After his engine froze due to an oil leak, he was seen to stand up in the cockpit preparatory to the jump, but just at that time the plane flipped over on its back and dove down. At about 100 feet above the water, Capt. Vetter did succeed in getting free of his plane and pulled his chute, but it was too late. As a matter of fact, in January VMF-211 lost more planes in operational accidents than in combat.

On **27th January** 24 Corsairs of VF-17, over 40 more of VMF-211, -212 and -321, and a dozen of New Zealand's Kittyhawks joined 24 Mitchells bound for Lakunai. The defending force was estimated at 60-70 fighters. The *Jolly Rogers* shot down 16 Zeros, but again one of them failed to return. Marine Corsair pilots were credited with seven Zeros, including three by Lt. Franklin Thomas of VMF-211.

VMF-321 contributed with two victories but nearly lost its only ace, Lt. Robert See. As the squadron's diary relates: "Lt. See's plane caught on fire under dashboard, from a short circuit, while over Rabaul area. He opened his hood, lost altitude, and the fire went out. He was therefore able to continue on course."

LtCdr Blackburn was alarmed by the loss rate in his squadron – three pilots in the first two days of their new tour. He came up with the idea of Roving High Cover, which he put to the test on **28th January**. Six Corsairs, flying ten minutes ahead of the main raid and as high as 30,000 feet, were to engage and, if possible, scatter any Zeros forming up after takeoff. On that day 20 Corsairs of VF-17, 16 of VMF-211 and 8 of VMF-215, together with 12 RNZAF Kittyhawks, 8 Hellcats and 12 Lightnings, escorted 48 SBDs and 18 TBFs on a strike mission to Tobera. On the way to the target six Corsairs led by LtCdr Hedrick throttled up and outdistanced the main formation, heading straight for Rabaul. As expected, they surprised some 50 Zeros at low

SECRET

Monthly Report of Destruction of Enemy Ships and Planes and Loss of Own Aircraft

ENEMY PLANES

Date	Name of Pilot	Type U.S. Plane	Enemy Plane Destroyed	Probable	Damaged	Where
2Jan44	Major T.V.Murto, Jr	F4U	1 Zeke	1 Zeke		in air
"	Lt. J.C.Hundley	"	1 Zeke			"
3Jan44	Major J.W.Ireland	"	1 Zeke			"
"	Major R.L.Hopkins	"	1 Zeke	1 Zeke		"
4Jan44	Lt. G.S.Langston	"	1 Zeke			"
6Jan44	Major J.D.Howard	"	1 Zeke			"
7Jan44	Lt. R.D.Rushlow	"	1 Zeke			"
"	Capt. A.R.Vetter	"	1 Zeke			"
"	Major R.L.Hopkins	"	1 Zeke			"
8Jan44	Lt. M.R.Tutton	"	2 Jakes			in water
"	Lt. N.R.Landon	"	1 Jake			"
9Jan44	Lt. F.C.Thomas,Jr	"	1 Tony			in air
"	" " "	"	1 Hamp	2 Zeke	1 Zeke	"
"	Capt. W.L.Beerman	"	1 Zeke			"
"	Lt. R.E.Lee	"	1 Zeke			"
14Jan44	Major R.A.Harvey	"	1 Zeke			"
"	Capt. H.V.Winfree,Jr	"	1 Zeke			"
"	Lt. T.C.Czarnecki	"	1 Zeke			"
15Jan44	Major J.W.Ireland	"	1/3 Zeke			"
"	Capt. J.A.Paradis,Jr	"	1/3 Zeke			"
"	Lt. J.J.Daly	"	1/3 Zeke			"
17Jan44	Lt. J.C.Thornton	"	1 Zeke		1 Zeke	"
"	Lt. R.L.Stigall	"	1 Zeke		1 Zeke	"
"	Capt. J.A.Paradis,Jr	"	1 Zeke		2 Zeke	"
"	Lt. F.C.Thomas, Jr	"	1 Zeke		2 Zeke	"
20Jan44	Major J.D.Howard	"	1 Zeke			"
"	Lt. T.C.Czarnecki	"	1 Zeke			"
"	Lt. C.J.Wheeler, Jr	"	1 Zeke			"
"	Lt. C.I.Cobb, Jr	"	1 Zeke			"
"	Lt. J.C.Hundley	"	2 Zeke			"
22Jan44	Major J.D.Howard	"	1 Zeke			"
"	Lt. J.C.Hundley	"	2 Zeke			"
"	Lt. E.H.McCaleb	"	1 Zeke			"
"	Lt. R.L.Stigall	"	1 Zeke			"
"	Lt. R. Ahern	"		1 Zeke		"
23Jan44	Capt. W.E.Hower	"	2 Zeke			"
"	Capt. W.B.Thomson	"	2 Zeke		1 Zeke	"
"	Lt. F.C.Thomas, Jr	"	1 Zeke			"
"	Major T.V.Murto, Jr	"	2 Zeke			"
"	Lt. R.L.Stigall	"	1 Zeke		1 Zeke	"
"	Major J.W.Ireland	"	3 Zeke	1 Zeke		"
"	Capt. N.K.Toerge,Jr	"		1 Zeke		"
"	Capt. W.L.Beerman	"	1 Zeke	2 Zeke		"
"	Lt. R. Ahern	"	1 Zeke	1 Zeke		"
"	Capt. H.C.Langenfeld	"	1 Zeke	1 Zeke		"
"	Lt. J.C.Thornton	"	1 Zeke			"
"	Lt. R.E.Lee	"	1 Zeke			"
24Jan44	Lt. H.E.Segal	"	2 Hamp			

SECRET 76

January 1944, combat-intense month for VMF-211 – 71 confirmed victories.

SECRET

ENEMY PLANES (cont.)

Date	Name of Pilot	Type U.S. Plane	Destroyed	Probable	Damaged	Where
24Jan44	Lt. F.C.Thomas, Jr	F4U		1/2 Zeke	1 Zeke	in air
"	Major R.L.Hopkins	"		1/2 Zeke		"
27Jan44	Lt. R.E.Lee	"	1 Zeke			"
"	Lt. J.C.Thornton	"	1 Zeke			"
"	Lt. F.C.Thomas, Jr	"	3 Zeke			"
28Jan44	Capt. H.V.Winfree,Jr	"	2 Zeke			"
"	Lt. H.W.Mosley	"	1 Zeke			"
"	Lt. E.H.McCaleb	"	1 Zeke			"
"	Lt. M.R.Tutton	"	1 Zeke			"
"	Lt. T.C.Czarnecki	"	1 Hamp			"
"	Lt. J.C.Hundley	"	1 Hamp			"
29Jan44	Capt. J.A.Paradis,Jr	"	2 Zeke			"
"	Lt. R.D.Rushlow	"	1 Zeke			"
"	Major J.W.Ireland	"	1 Zeke			"
"	Lt. F.C.Thomas,Jr	"	1 Zeke			"
"	Lt. E.G.Nelson	"	1 Zeke			"
"	Capt. W.L.Beerman	"		1/2 Zeke		"
"	Lt. R.E.Lee	"		1/2 Zeke		"
30Jan44	Lt. G.S.Langston	"	1 Zeke			"
31Jan44	Lt. R.D.Rushlow	"	1 Zeke			"
"	Lt. F.C.Thomas,Jr	"	1 Zeke			"
"	Major J.W.Ireland	"		1 Zeke		"
"	Lt. R. Ahern	"		1 Zeke		"
"	Lt. E.G.Nelson	"			1 Zeke	"

ENEMY SHIPS

Date	Type of ship	extent of damage	No. & type of planes	Comment
8Jan44	Barge (150 ft.)	Burned	2 F4U	Pilots: Major W.T.Campbell and Lt. H.E.Segal strafed and set fire to both barges, hitting numerous personnel aboard.
"	Barge	"	2 F4U	

FRIENDLY LOSSES

Date	type of plane	cause	status of crew	Comment
8Jan44	F4U	Missing	M.I.A.	Did not return to base
13Jan44	F4U	Crashed at sea	Killed	Shortland area
14Jan44	F4U	Missing	M.I.A.	Over Rabaul
18Jan44	F4U	Crashed on landing	Slight injuries	
22Jan44	F4U	" "	" "	
26Jan44	F4U	Crashed at sea	Uninjured	Empress Augusta Bay
30Jan44	F4U	Midair collision	Killed	Over base

SECRET

77

level, and made the most of the opportunity. Five VF-17 pilots – Ens. Percy Divenney, Lt(jg) Douglas Gutenkunst, Lt(jg) Paul Cordray, Lt. Harry March and Lt(jg) Tom Killefer – each got two. Others contributed with more, for a total of 14.5 (one shared with VMF-215), for no losses of their own.

VMF-215 bagged 5.5 Zekes (one shared with VF-17). Of these, four fell to the guns of Capt. Donald Aldrich; he came back with shrapnel wounds, and his Corsair was so badly shot up that it was written off. VMF-211 contributed with seven victories (two by Capt. Harry Winfree).

On the same day a new Corsair squadron, VMF-217 *Wild Hares*, arrived on Bougainville, relieving VMF-321.

On **29th January** *Jolly Rogers*, VMF-211 and -212 again escorted SBDs and TBFs to Tobera. The frantic pace of operations reduced the number of VF-17's serviceable machines down to 20, of which four turned back after takeoff. Therefore, only two Corsairs were left to fly as the Roving High Cover. Their pilots, Lt(jg) Ira Kepford and Lt(jg) Howard Burriss, were nevertheless positive that they could handle the situation. As the two arrived over Rabaul at 30,000 feet, they spotted 12 Zeros circling at 24,000 feet over Cape Gazelle. For the next ten minutes Kepford and Burriss took turns in hitting the hapless Japanese, pouncing on them from above in high stern runs and zooming up to regain altitude. Each of them shot down four, raising Kepford's tally to 11, and Burriss's to 7.5. The approaching main bomber formation and their escorts could see in the distance eight flaming torches falling down one after another into the ocean.

The Japanese fighters that chose to mix it up were tied down by pilots of VMF-211 and -212

who scored 6 and 5 victories, respectively. The remaining VF-17 pilots, who accompanied the bombers, chipped in two more. Hence, the box score of AirSols Corsairs was 21.

On **30th January** *Jolly Rogers* flew cover for B-25s targeting a supply dump near Lakunai. Opposition in the air was practically non-existent, and VF-17 scored two victories only because one flight ranged out ahead of the main formation. VMF-215 didn't even engage, but somehow lost a pilot. VMF-217 pilots flew their combat debut, which proved quite uneventful.

In the late afternoon the AirSols HQ raised alarm – an allied long-range reconnaissance aircraft had reported seeing a Japanese carrier making for Rabaul. It sounded too good to be true, but HQ was taking no chances, and ordered every available aircraft to strike. The escort for 16 SBDs and 18 torpedo-armed TBFs was provided by 47 Corsairs (of VF-17, VMF-211, -215 and -217) and 8 RNZAF Kittyhawks. There was no carrier to be seen in the vicinity of Rabaul, but this time the defenders swarmed up in force – some 50-60 fighters were scrambled to contest this incursion.

VMF-215 pilots scored 11 victories, including two by Lt. Chandler and four by Lt. Hanson (who identified his opponents as two Zekes and two Tojos). *Jolly Rogers* came back home with ten victories – 'doubles' by LtCdr Blackburn, Lt. Oscar Chenoweth, Lt(jg) Ira Kepford and Lt(jg) Merl Davenport – but without one pilot, most probably shot down in the massive melee over Rabaul.

As dozens of aircraft began to streak back from this impromptu raid, chaos broke out at the Bougainville airfields. In the fading light ground controllers at both Piva airstrips quickly lost control of the situation. Battle-damaged aircraft were landing haphazardly, disregarding any traffic pattern. LtCdr Blackburn squeezed in at Piva Uncle, but as his wingman lined up to land he collided in mid-air with another Corsair. Both pilots, Lt(jg) Gutenkunst and Maj. Hopkins of VMF-211, perished in a ball of fire (at the time of his death Gutenkunst had four victories to his

Lt(jg) Tom Killefer of VF-17, Green Islands, March 1944.

F4U-1A of VMF-223 undergoing repairs; Green Islands, spring 1944. Note uncovered main fuel tank, seen directly ahead of the windshield. Apparently a shell splinter hit the aircraft's starboard side and pierced the tank.

credit). Moreover, in a series of hair-raising crash landings VF-17 lost three more Corsairs, but their pilots survived.

At the end of this intense day three Corsair squadrons – VF-17, VMF-215 and VMF-211 – were credited with a total of 24 victories (12, 11 and 1, respectively).

On **31st January** VF-17 could field only 16 airworthy Corsairs – too few to provide a Roving High Cover. It meant that this time the enemy would have the advantage of altitude, and very likely make the most of it. The target for a bunch of SBDs and TBFs was Japanese shipping in Karavia Bay. The bombers and their escorts arrived at 14-15,000 feet. The Zeros, orbiting as usual at about 25,000 feet, held the initiative. A single Zero made a pass, hitting Lt(jg) Burriss as he went. The stricken Corsair floated down belching smoke. The subsequent fate of Howard Burriss, an ace with 7.5 victories, remains unknown. In retrospect, January 1944 imposed a heavy strain on the *Jolly Rogers*. Their victory tally amounted to 60.5 'kills' in just six days, but at the price of 6 pilots and 13 aircraft.

On the same day 20 Corsairs of VMF-212 and -217, supported by a dozen of RNZAF Kittyhawks, escorted 24 SBDs to Tobera airfield, running into some 20-30 enemy fighters. Lt. Louis Russell of VMF-217 scored the first victory for his squadron. VMF-212 tallied three victories, but lost a pilot. It *nearly* lost another, Lt. G.E. McClane, who scored a victory under somewhat dramatic circumstances:

"A Zeke got on my tail and filled me full of holes before I had a chance to begin to maneuver. One 20 mm exploded my ammunition and set my left wing on fire. I figured I was finished and when I saw a Zeke on my level, 15,000', I just eased up on his tail and started firing. He made no evasive maneuver. Just rolled over and flamed down. When I finished the attack I saw the fire on my wing had gone out. I was in no shape to stick around so I headed for home."

The squadron's diary also includes this interesting observation on enemy tactics: "Again phosphorous bombs were dropped by early interceptors. It is commented on again that these are used as vectoring signals to other enemy fighters rather than in an effort to damage our fighters. [The bursts] can be seen from a great distance and hold for some time."

On **1st February** VMF-211, stationed at Piva Yoke, ended its very intense combat tour and was relieved by VMF-222.

Heavy air strikes on Rabaul continued on **3rd February**. The opposing force was estimated at only 15 fighters. VMF-212, -215, -217 and -218 (the latter squadron flying its combat debut) collectively bagged seven victories, including two by Capt. Harold Spears of VMF-215. On the return leg Lt. Robert Hanson of VMF-215 was killed in action. Against orders he engaged in a duel with an antiaircraft battery at Cape St. George, at the southernmost tip of New Ireland. He was hit, one wing ripped away, and the aircraft plunged into the sea. At the time of his death Hanson was credited with 25 victories on Corsairs – more than any other Navy or Marine pilot.

4th February was a bad day for VF-17, which escorted Marine B-24s to Tobera. About 30 Japanese fighters took off to oppose the raid. As some of the Zeros closed in, Ens. Divenney reached for the valve which would purge the gasoline vapor from his wing tanks with CO_2. Instead, he inadvertently activated the landing gear emergency release system (the two levers differed only in color). With his wheels stuck in the lowered position, he was unable to keep up with the rest of the flight. LtCdr Blackburn immediately ordered him to move under the cover of the bomber gunners, but it was already too late – Divenney was shot down. A few minutes later Lt(jg) Malone began to lag behind his division. Again, a vigilant Zero pilot flamed him in a single, slashing attack.

Clear weather on **6th February** enabled the AirSols to mount a major raid, once more against Lakunai. The striking force consisted of three B-25 squadrons. On their way to the target the Mitchells rendezvoused with their escorts: 40 Corsairs, 24 'Kiwi' Kittyhawks and 8 Hellcats. LtCdr Blackburn rushed ahead with one division, climbing at full throttle. When the four Corsairs of VF-17 arrived over Rabaul at 25,000 feet, their pilots were greeted with the welcome sight of about 40 Zeros forming up at 12,000 feet over Tobera and Rapopo. Out of the sun came the four Corsairs of the Roving High Cover, their guns ablaze. After every pass they would zoom up, wing over and strike again. Blackburn knocked down four (they were his last victories, giving him an ultimate tally of 11). Lt(jg) Robert Mims, the leader of the division's second section, flamed three fighters (including one he recognized as a Ki-44).

On **7th February** 60 Navy and Marine SBDs set out to strike Vunakanau; 40 Corsairs of VMF-215, -218 and -222 tagged along, as well as 12 RNZAF Kittyhawks. Some 30-35 Zekes, Hamps and Tonys accepted the challenge, and paid the price. Pilots of VMF-215 shot down eight fighters over Rabaul (Capt. Aldrich, Lt. Herman and Lt. Williams bagged two apiece) for no losses to themselves.

LtCdr Blackburn had something up his sleeve for the Japanese on that day. On the return leg, while the bombers continued to Bougainville, the six Corsairs he was leading made a wide orbit to the north, and went flat out, low over water, along the eastern coast of New Ireland. After some time they swerved to port and raced across St George's Channel. They roared across Lakunai airfield at rooftop height, catching the Japanese off-guard. Numerous Zeros were encountered in the landing pattern or taxing about. A few of them, raked with long bursts of fire, were claimed damaged or probably destroyed – there was no time to look back to confirm their fate. One Zero, which happened to be pulling up after an aborted landing, crashed onto the runway in a shower of flames. Two of the three confirmed victories were credited to Lt(jg) William Landreth.

On **9th February** Vunakanau was again on the receiving end. The strike force consisted of 24 B-25s. Their escorts – some 60 Corsairs of VMF-215, VMF-217 and VF-17 – were more than enough to keep at bay 15-20 Zeros that were in the area. The box score for the three Corsair squadrons was 10 victories; Lt. Edwin Hernan of VMF-215 got three, while Capt. John Hench and Lt. William White of VMF-217 bagged two apiece.

On **10th February** AirSols launched a series of bombing raids against Rabaul airfields. Zero pilots seemed little enthusiastic about mixing it up with Corsairs. Nevertheless, the attrition continued – VF-17 scored two victories; three were credited to the *Wild Hares* (VMF-217), of which Capt. William Peek tallied two (the latter squad-

F4U-1A of VMF-223 and a C-47 on Green Island, March 1944.

ron joined the air battle over Rabaul too late to amass a large tally; nonetheless, by the end of the month it had 16 victories to its credit for the loss of one pilot).

On **12th February**, in very much the same circumstances, VMF-215 bagged two, as did VMF-217 (both were credited to Maj. John Bohnet). VF-17 escorted B-25s to Tobera and didn't even see enemy fighters.

On **13th February** Capt. William Carlton, one of the aces with VMF-212, shot down three Zeros in the vicinity of Tobera. Maj. George Poske of the same squadron 'made ace', knocking down a Zero over Cape Gazelle. The 253rd Kokutai lost one of its aces, *Hiko Heisocho* Nobuo Ogiya, during an attack on some SBDs and their escorting Corsairs; he had claimed a record 18 victories in just 13 days before his death.

The following day VMF-215, having completed its third and last tour in the Solomons, left for Guadalcanal, and later for Sydney. Once all the submitted claims had been evaluated and credits allowed, VMF-215 emerged as the top-scoring Marine squadron of the Solomons campaign, with 135.5 victories.

On **15th February** the New Zealanders landed unopposed on the Green Islands, a small atoll located less than 150 miles from Rabaul. On that day Vice Admiral Kusaka once again launched his bombers. VMF-212 intercepted 15 Vals and shot down six of them. Three fell to Lt. Philip DeLong, making him a 'double ace'. Most interestingly, these were DeLong's last victories of the war, but not the last ones in his Corsair pilot career.[12] On the same day 2/Lt. Charles Jones, one of the aces with VMF-222, scored his last victories – two Zeros in Rabaul area.

Obviously, attrition worked both ways. On **17th February**, while escorting a bunch of SBDs hunting for ships in Simpson Harbor, the *Jolly Rogers* came across 30-40 fighters. The Japanese attacked with a breathtaking vigor, scattering the Corsairs. Ens. Dunn claimed a Zero, but seconds later he was separated from his wingman and went missing. Lt(jg) Miller's aircraft was hit by an anti-aircraft round, which blasted off part of the starboard wing. Miller bailed out of the spinning Corsair and was taken prisoner; he perished in captivity.

On the same day Maj. Donald Sapp of VMF-222 shot down two Zeros over Rabaul.

The following day, **18th February**, the *Jolly Rogers* returned with a vengeance. Sixteen Corsairs accompanied some B-24s. As they slowly droned on to Vunakanau, LtCdr Roger Hedrick sped ahead with two divisions flying as the Roving High Cover. The eight Corsairs turned up over Rabaul half an hour before the bombers, poised at 32,000 feet. As small groups of Japanese fighters began to assemble after takeoff, Hedrick and his pilots bounced them time and again. They shot down seven, including three by Hedrick, and two by Lt(jg) Earl May (yet another ace of VF-17).

The last battle to feature Corsairs against regular IJNAF units over Rabaul took place on **19th February**. On that day, *Hiko Heisocho* Tetsuzo Iwamoto led aloft 26 Zeros of the 253rd Kokutai. The *Jolly Rogers* again split up their forces. While 22 machines guarded the SBDs and TBFs making for Lakunai, Lt. Oscar Chenoweth and his wingman Lt(jg) Daniel Cunningham outpaced them by some 20 minutes. The Zeros were circling lazily at 18,000 feet, waiting for the bombers to arrive. With over 10,000 feet of height advantage on their side, Chenoweth and Cunningham went down like screaming banshees. Yet again the Japanese found themselves powerless against the high-sided attacks, repeated with infuriating impunity. The results were telling – Cunningham

F4U-1A Corsair of VMF-222 on the ground being maintained by Marines; Bougainville, April 1944.

F4U-1A issued to No 20 Fighter Squadron RNZAF in April 1944, and transferred to No 23 Squadron by November 1944.

added four Zeros to his former tally of three; Chenoweth bagged three (one reported as a Tojo) to raise his final score to 8.5.

The Japanese fighters reformed and engaged the escorting Corsairs as the bombers were turning for home. The *Jolly Rogers* were ready. They were credited with six more Zeros, including three by Lt(jg) Earl May. Meanwhile Lt(jg) Kepford, the squadron's highest-ranking ace, flying by himself after his wingman had to turn back, spotted a lone Rufe off Cape Siar (at the southern tip of New Ireland), and splashed it in one pass. The next moment he was bounced by four Zeros. He shot down one, made another crash, and outran the remaining two, returning to Piva Yoke after four hours and 20 minutes – too exhausted to climb out of the cockpit on his own. He was credited with three victories (for a total of 16). In what turned out to be their last aerial battle, the *Jolly Rogers* had racked up 16, for no losses of their own.

The fate of the Bismarck Barrier had been sealed elsewhere. For the US Navy fast carrier groups roaming freely in the open ocean, Rabaul was not an obstacle. On 17th and 18th February carrier-borne aircraft struck at Truk, the main anchorage of the Japanese Navy in the South Pacific. Bypassed in that fashion, Rabaul became a useless stronghold. On **20th February** Admiral Koga ordered the remaining airworthy aircraft at Rabaul to withdraw to Truk, and the 11th Air Fleet was disbanded.

Rabaul had truly proved a 'hornets' nest'. One of the first squadrons to learn it the hard way was VMF-214 *Black Sheep*. This crack unit lost eight pilots, including two Majors – the CO and his second-in-command – in less than two weeks. VF-17 *Jolly Rogers* joined the offensive against Rabaul in late January, and lost ten pilots, including eight shot down by Japanese fighters. The remaining squadrons didn't fare much better. Altogether, 39 Corsair pilots were killed or went missing during Rabaul missions in January and February 1944.

The importance of the allied victory in the air campaign over the upper Solomons and Rabaul lay not in breaking through the Bismarck Barrier, which had lost its strategic significance well before this campaign was over. It was, almost from the start, a battle of attrition. This drawn-out slugfest proved costly for both sides, but for the Imperial Navy it was a disaster with far-reaching consequences. In order to defend the Barrier, the Japanese had stripped their last carriers of their air groups. The once-formidable elite of the IJNAF fighters had been thrown away on a series of wasteful operations. They were replaced by dozens of ill-trained tyros, who were later decimated by the Navy fighters in the 'Great Marianas Turkey Shoot' and similar engagements.

The turn of January and February 1944, the apex of the aerial battle over Rabaul, was also the most successful period for VMF-215 *Fighting Corsairs*. Capt. Arthur Warner, a veteran of three

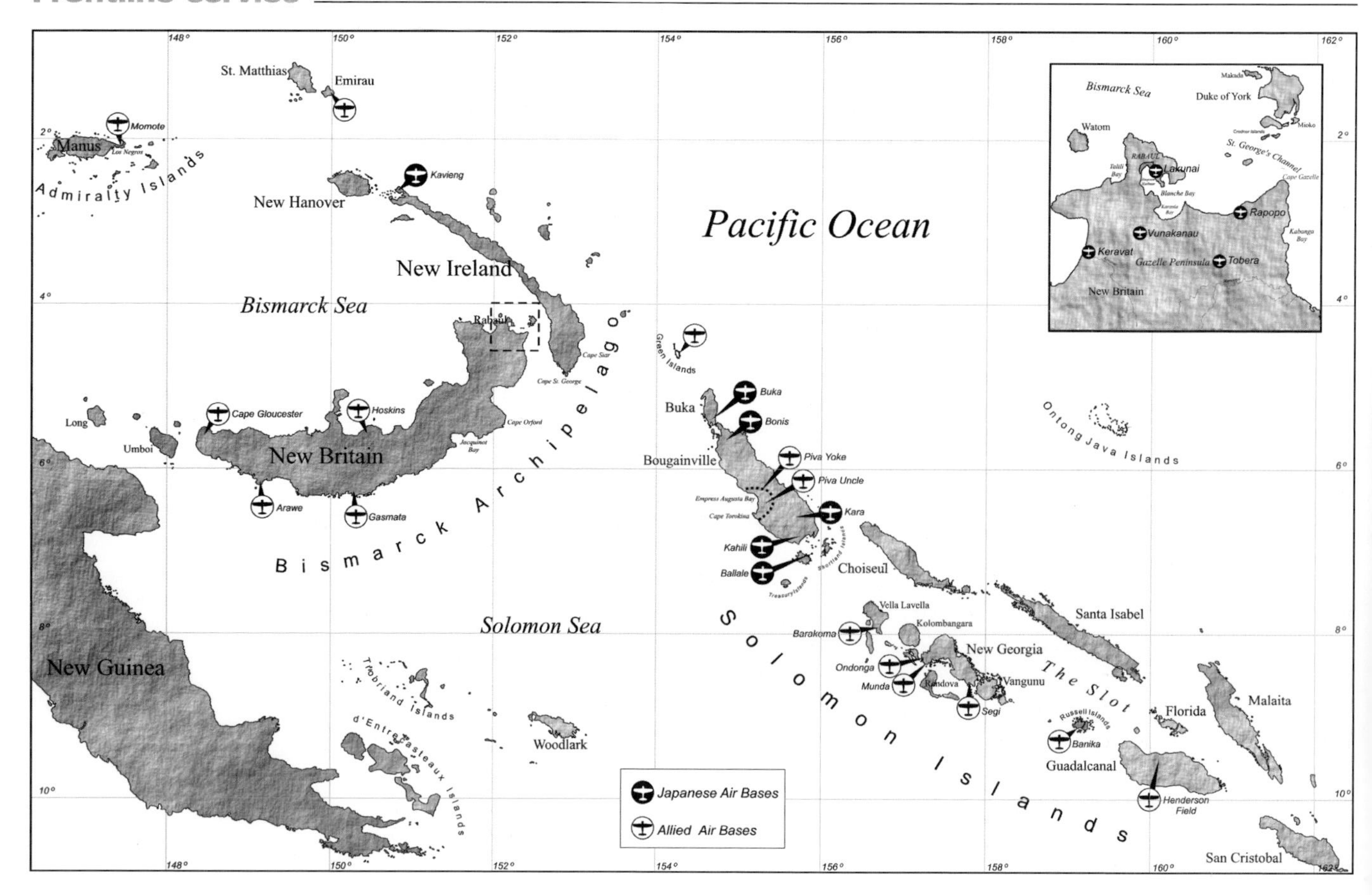

The Solomons and the Bismarck Barrier at the turn of 1943/44 (by Janusz Światłoń).

tours in the Solomons, had claimed his first victory over Empress Augusta Bay on 1st November 1943; by 12th February 1944 he had added seven more. Lt. Edwin Hernan, who had flamed his first on 18th January, had eight victories under his belt by 9th February. In the same period Capt. Harold Spears added 11 to his previous total of four. 2/Lt. Gerard Williams tallied seven Zeros in Rabaul area in less than three weeks. Lt. Creighton Chandler flew four sorties to Rabaul in a dozen or so days, and scored six victories (a Tony and five Zeros). Capt. Richard Braun flew to Rabaul five times in January 1944, and each time he shot down one Zero.

But even such successes as these paled in comparison with the achievements of the squadron's two leading aces. When Donald Aldrich offered his services to the Army Air Corps (he had a civilian pilot's license), he was turned down flat only because he had recently married. Trained by the RCAF as a fighter pilot, he joined the USMC in 1942. Subsequently, Capt. Aldrich served three tours with VMF-215, amassing 20 victories (all on Corsairs). Of that total, as many as 15 were scored over Rabaul, between 12th January and 9th February 1944.

There was only one pilot who fared even better. Lt. Robert Hanson, the top-scoring Corsair ace of the Second World War, bagged 20 in two weeks! As fate would have it, he fell victim to antiaircraft fire. He was posthumously awarded the Congressional Medal of Honor. Lt.Col. Harold A. Langstaff, who served with VMF-215, recalled:

"We went out to the Solomons with about twenty-eight pilots, and I think twelve didn't come back. We lost two in one night during a predawn takeoff. With no horizon, you had to go on instruments as soon as you left the ground. One of our pilots took off and spun in immediately off the end of the runway. The second plane taking off – we figure he saw the explosion and was looking down at it as he took off – developed vertigo, and spun in right on top of it. So we lost two planes and two pilots just like that.

We lost two more pilots who tried to destroy an antiaircraft battery. There was one island that had been bypassed, and everybody was warned to stay away from it because there was a Japanese antiaircraft battery that was deadly. On two different occasions two of our pilots tried to put that battery out of action and were lost as a result. In fact, Maj. Tomes, who was my flight leader, was one of them. He was on his way back from a strike and elected to try and knock out that gun and was killed. Then there was our Medal of Honor winner, Bob Hanson – he was killed trying the same thing".[13]

Individual achievements by other Corsair squadrons, although somewhat overshadowed by the successes of the *Black Sheep*, the *Jolly Rogers* and the *Fighting Corsairs*, are by all means noteworthy. Lts. Franklin Thomas and John Hundley, two aces with VMF-211 and veterans of three tours in the Solomons, scored all their victories (Thomas nine, Hudley six) in Rabaul area in January 1944. Also Maj. Hugh Elwood, the commander of VMF-212, became an ace in just three sorties to Rabaul in January 1944.

In VMF-321 *Hell's Angels* the only pilot to collect five victories at the controls of a Corsair was Lt. Robert See. He shot down five Zeros in the area of Rabaul at the turn of 1943/44. In VMF-222 the unquestionable leader in the scoring stakes was Maj. Donald Sapp, another veteran of three tours in the Solomons. In late 1943 he bagged four kills in Bougainville area (including a Ki-49 Helen bomber, a rare prey). In February and March 1944 he was credited with six Zeros in scraps over Rabaul.

VF-17 *Jolly Rogers*, 'orphaned' by the Navy and banished to fight among the Marines as a land-based unit, had a fair chance of becoming a forgotten squadron with broken morale. Far from that, they amassed an unparalleled 152 confirmed victories in just 76 days. In fact, they outscored all other allied fighter squadrons participating in the Solomons campaign and the subsequent siege of Rabaul. As many as 13 aces, each with at least five victories, served with the Fighting-17.

After the memorable scrap on 19th February the *Jolly Rogers* had no opportunity to meet the enemy in the air for the rest of their tour. Nonetheless, they found a new way to harass Rabaul. Shortly before the squadron's return to the USA in early March, some of VF-17's Corsairs were experimentally fitted with centerline bomb racks. It was a portent of things to come, for with time Corsair became a capable fighter-bomber.

Meanwhile, in mid-February VMF-223 returned for another tour of duty in the Solomons. Somehow it missed the last big fight over Rabaul on the 19th. On **25th February** two of the squadron's Corsairs led by Harlan Stewart, now a Major, took off on a strafing search for the float plane that had been sighted in Buka Passage. Maj. Stewart was shot down and killed by AA fire. On **27th February** Capt. Fred Gutt scored his eighth and last victory, shooting down a Rufe near Borpop airfield on New Ireland.

On **29th February** the Americans landed on the Admiralty Islands, and a month later they made it to Emirau, cutting off the Bismarck Barrier from the rear. Nevertheless, keeping Rabaul in check was a daunting task. Over the ensuing months many allied airmen fell foul of the stronghold's anti-aircraft guns.

After the IJNAF regular units had withdrawn from Rabaul, some 30 Japanese fighter pilots, most of them sick with malaria at the time of the retreat, were left behind. Together with a group of mechanics, stranded at Rabaul due to lack of transport, they organized an improvised fighter unit. At the turn of February and March they assembled eight airworthy Zeros from numerous wrecks. A clash with regularly patrolling Corsairs was only a matter of time.

On **3rd March** VMF-223 pilots came across seven Zeros in the vicinity of Tobera airfield. Maj. Robert Keller, the squadron's CO, shot down one of them. On **12th March** Corsairs of VMF-222 encountered seven Zeros orbiting over Tobera. The Japanese sought protection under the fire of the airfield's anti-aircraft guns, circling at barely 300 feet. Maj. Donald Sapp (the executive officer and the squadron's leading ace) and Capt. Robert Wilson dived into the anti-aircraft barrage and shot down three Zeros: Sapp two, and Wilson one. Meanwhile, three Zeros slipped away to the north, only to run into the Corsairs of VMF-216; Lt. Jean Patton bagged one.

Lt. Roland Heilman of VMF-222 won the last victory of the war to involve Corsairs and Zeros over the Bismarck Barrier. On **13th June** 1944 two Zeros were dispatched from Rabaul to reconnoiter the allied fleet at Los Negros anchorage in the Admiralty Islands (at that time the Japanese garrison in Rabaul was still convinced that there was a real threat of an impending invasion). When a storm forced the two Zeros to turn back, one of them winged its way to Kavieng. There, over the airfield, it was jumped and shot down in flames by Lt. Heilman on a routine patrol.

The last Corsairs to continue the unrewarding task of isolating Rabaul were those of RNZAF squadrons. Some of the 'Kiwi' units were veteran outfits that had participated in earlier battles in the area. They had exchanged their aging Kittyhawks for F4Us. The first of them, No 20 Sqn RNZAF, joined the Americans on Bougainville in mid-May 1944. At the turn of 1944/45 the New Zealanders took over responsibility for the area from the Marine squadrons, which were soon to join the battle for the Philippines (VMF-115, -211 and -313 left Emirau in December 1944; VMF-218 moved out of Green Islands in November 1944, and VMF-222 in January 1945). By the end of the war there were nine RNZAF Corsair squadrons stationed around the Bismarck Barrier: four on Bougainville, two at Los Negros (Admiralty Islands), and three on New Britain.

Bibliography:

Blackburn Tom / Eric Hammel, *The Jolly Rogers*, St Paul 2006.
Bowman Martin W., *Vought F4U Corsair*, Ramsbury 2002.
Boyington Gregory, *Baa Baa Black Sheep*, New York 1958.
Carl Marion E./Barrett Tillman, *Pushing the Envelope. The Career of Fighter Ace and Test Pilot Marion Carl*, Annapolis 2005.
Cox Bryan, *Too Young To Die – The Story of a New Zealand Fighter Pilot in the Pacific War*, Ames 1989.
Cook Lee, *The Skull & Crossbones Squadron: VF 17 in World War II*, Atglen 1998.
DeBlanc Jefferson J., *The Guadalcanal Air War*, Gretna 2008.
Gamble Bruce D., *The Black Sheep*, New York 2000.

Hammel Eric, *Air War Pacific: Chronology*, Pacifica 1998.
Olynyk Frank, *Stars & Bars – A Tribute To The American Fighter Ace 1920-1973*, London 1995.
Petty Bruce M., *At War in the Pacific: Personal Accounts of World War II Navy and Marine Corps Officers*, Jefferson 2006.
Reinburg Hunter, *Combat Aerial Escapades*, New York 1966.
Sakaida Henry, *The Siege of Rabaul*, St Paul 1996.
Sakaida Henry, *Imperial Japanese Navy Aces 1937-45*, Oxford 1998.
Tillman Barrett, *Corsair: The F4U in World War II and Korea*, Annapolis 2002.
Toliver Raymond F. / Constable Trevor J., *Fighter Aces of the U.S.A.*, Fallbrook 1979.
Walton Frank, *Once They Were Eagles*, Lexington 1996.
War Diaries of: VMF-112, -121, -122, -123, -124, -211, -212, -213, -214, -215, -216, -217, -221, -222, -223, -321; memorandum on VF(N)-75; National Archives and Records Administration, Washington.

Endnotes

[1] Jefferson J. DeBlanc, *The Guadalcanal Air War*.
[2] Bruce M. Petty, *At War in the Pacific*.
[3] All quotes, unless otherwise noted, come from relevant squadron war diaries.
[4] Of note is a tremendous success by Guadalcanal-stationed VF-21, a Wildcat unit of the Navy, which scored 30 victories on that day.
[5] On 7th April 1943 James Swett, at that time still flying Wildcats, had knocked down seven Vals in one sortie – a feat which earned him the Medal of Honor.
[6] F. Walton, *Once They Were Eagles: The Men of the Black Sheep Squadron*. Walton was the squadron's intelligence officer.
[7] The Marines also had their night fighter squadron in the Solomons – VMF(N)-531 – but it operated twin-engined PV-1N Venturas.
[8] H. Sakaida, *The Siege of Rabaul*.
[9] (Maj.Gen.) Marion Carl / B. Tillman, *Pushing the Envelope*.
[10] This tremendously successful squadron tallied 111.5 victories during its first tour on Guadalcanal in 1942, flying F4F Wildcats.
[11] R.F. Toliver / T.J. Constable, *Fighter Aces of the U.S.A.*
[12] On 21st April 1951, whilst flying an F4U-4, Capt. DeLong shot down two Yak-9 fighters over Korea, which gave him a total score of 12 individual and three shared victories.
[13] Bruce M. Petty, *At War in the Pacific*.

Appendixes

Operational characteristics and performance – F4U-1D/FG-1D

Loading condition		Combat (1)	Combat (2)	Combat (3)	Fighter (4)	Fighter (5)	Bomber (6)	Bomber (7)	Rocket (8)
Power condition*		combat	maximum	normal	normal	normal	normal	normal	normal
Weights and external stores:									
Max. takeoff weight	[kg]	5,523	5,523	5,523	6,017	6,518	6,468	6,428	6,539
Oil	[liter]	49	49	49	83	91	83	49	83
Auxiliary fuel tanks	[liter]	-	-	-	1×568	2×568	1×568	-	1×568
Bomb load	[kg]	-	-	-	-	-	1×454	2×454	-
Rockets (number × caliber)	[mm]	-	-	-	-	-	-	-	8×127
Performance:									
Max. speed at a given altitude	[kph]	658	637	634	602	568	568	568	562
	[m]	6,066	7,285	7,559	7,498	7,376	7,376	7,376	7,376
Max. speed at sea level	[kph]	576	552	516	491	467	467	467	460
Initial climb rate	[m/s]	17.1	14.9	11.6	10.2	9.0	9.0	9.1	8.8
Time to climb to 3,048/6,096 m	[min.]	3.3/7.1	3.8/8.3	4.6/9.7	5.2/11.2	6.0/13.1	6.0/13.1	6.0/13.1	6.1/13.2
Service ceiling	[m]	12,192	12,131	11,979	11,521	11,125	11,125	11,125	11,034
Combat radius**	[km]	-	-	185	639	1,028	611	157	583
Max. range at a given speed***	[km]	-	-	1,577	2,414	3,049	2,156	1,295	2,124
	[kph]			293	283	283	283	283	283
Takeoff run****	[m]	199/86	199/86	199/86	256/116	337/160	329/156	322/153	341/164

Notes:
– Performance is based on flight test.
– In each condition: max. amount of ammunition (2,400 rounds), max. amount of fuel in the internal tanks (897 liters/237 gal), two underwing pylons (capped when not in use) and one centerline rack, empty weight 4,089 kg.
– Underwing rocket launchers only in condition no. 8.
– Condition no. 5 – one of the wing droppable tanks is self-sealing and is carried the entire distance.
– In clean condition (as in condition no. 1, but without underwing pylons and centerline rack): max. speed 589 kph at sea level and 671 kph at 6,096 m (20,000 ft).
– In ferry condition (armament removed, max. takeoff weight 5,945 kg, amount of fuel held internally 897 liters, auxiliary tanks 2×568 liters/150 gal): max. range 3,186 km (at 283 kph, at 457 m); addition of under-fuselage droppable fuel tank (662 liters/175 gal) increases max. range to 4,039 km.
– Condition no. 4 with 662-liter fuel tank under fuselage in lieu of 568-liter underwing tank: max. range 2,623 km (at 283 kph, at 457 m), max. speed 497 kph at sea level.
– In condition with empty rocket launchers (as in condition no. 1, but with rocket launchers fitted): max. speed 563 mph at sea level and 664 kph at 6,066 m (19,900 ft).

* Compare R-2800-8/-8W power rating table; ** Fighter combat radius formula: (1) Warm-up 20 min., (2) Takeoff 1 min., (3) Rendezvous 20 min. at sea level at 60% Normal Power, (4) Climb to 4,572 m (15,000 ft) at 60% Normal Power, (5) Cruise out at 4,572 m (15,000 ft) at velocity for max. range, (6) Drop tanks and bombs, fire rockets, (7) Combat 20 min. at 4,572 m (15,000 ft), 8.5 min. WEP and 11.2 min. Military Power and descend, (8) Cruise back at 457 m (1,500 ft) at 315 kph TAS, (9) Reserve 60 min. at velocity for max. range. Radius = climb + cruise out = cruise back; *** At 457 m (1,500 ft); **** Calm/head-on wind 46.3 kph (25 knots).

Source: *Airplane Characteristics & Performance, Model F4U-1D, -1C also FG-1D*, Bureau of Aeronautics, Navy Department, 1 August 1945.

Production of F4U-1/F3A-1/FG-1 Corsair (split into producers and models)

Model	Total number	Number in BuNo block	Bureau Numbers (BuNo)	Notes
Vought-Sikorsky Aircraft Division of United Aircraft Corp., Stratford, Connecticut, USA (until 1943) Chance Vought Aircraft Division of United Aircraft Corp., Stratford, Connecticut, USA (since 1943)				
XF4U-1	1	1	1443	prototype
F4U-1	758	584	02153–02736	1 converted to XF4U-2, 25 converted to F4U-2, 1 converted to XF4U-3A
		40	03802–03841	3 converted to F4U-2
		64	17392–17455	3 converted to F4U-2
		70	18122–18191	
F4U-1A	2,056	666	17456–18121	1 converted to XF4U-3A
		690	49660–50349	2 converted to F4U-2, 1 converted to XF4U-3B, 2 converted to F4U-4X
		700	55784–56483	
F4U-1C	200	3	57657–57659	produced concurrently with F4U-1D
		15	57777–57791	
		18	57966–57983	
		12	82178–82189	
		30	82260–82289	
		25	82370–82394	
		25	82435–82459	
		43	82540–82582	
		7	82633–82639	
		22	82740–82761	
F4U-1D	1,685	310	50350–50659	produced concurrently with F4U-1C
		573	57084–57656	
		117	57660–57776	
		174	57792–57965	
		70	82190–82259	
		80	82290–82369	
		40	82395–82434	
		80	82460–82539	
		50	82583–82632	
		100	82640–82739	
		91	82762–82852	
	[2]	[2]	82853, 82854	cancelled
XF4U-2	(1)	(1)	02153	
F4U-2	(33)	(31)	02243, 02421, 02432, 02434, 02436, 02441, 02534, 02617, 02622, 02624, 02627, 02632, 02641, 02672, 02673, 02677, 02681, 02682, 02688, 02692, 02708–02710, 02731, 02733, 03811, 03814, 03816, 17412, 17418, 17423	converted from F4U-1 at NAF
		(2)	49858, 49914	converted from F4U-1A in VMF(N)-532
XF4U-3A	(2)	(1)	17516	converted from F4U-1A
XF4U-3B		(1)	49664	
XF4U-3C		[(1)]	02157	planned converted from F4U-1, but was lost in a crash
Total Vought	4,700			
Brewster Aeronautical Corp., Johnsville, Pennsylvania, USA				
F3A-1	735	260	04515–04774	
		248	08550–08797	
		227	11067–11293	
	[773]	[353]	11294–11646	order cancelled
		[420]	48940–49359	
Total Brewster	735			
Goodyear Aircraft Co., Akron, Ohio, USA				
FG-1/-1A	2,010	2000	12992–14991	8 converted to XF2G-1
		10	76139–76148	
FG-1D	1,997	45	67055–67099	
		591	76149–76739	1 converted to FG-3
		666	87788–88453	
		695	92007–92701	25 converted to FG-3
	[755]	[155]	67100–67254	order cancelled
		[600]	92702–93301	
FG-3	(26)	(26)	76450, 92252, 92253, 92283, 92284, 92300, 92328, 92232, 92338, 92341, 92344, 92345, 92354, 92359, 92361, 92363, 92364, 92367, 92369, 92382–92385, 92429, 92430, 92440	converted from FG-1D (probably only some were completed)
Total Goodyear	4,007			
Grand Total	**9,442**			

Source: R.A. Grossnick, *United States Naval Aviation, 1910–1945*, Washington 1997, Appendix 9: *Bureau (Serial) Numbers of Naval Aircraft*; FAOW no. 88; B. Tillman, *Vought F4U Corsair*, North Branch 1996.

Deliveries of F4U-1/F3A-1/FG-1 Corsair (split into producers and years)

Producer	1940	1942	1943	1944	1945	Total
Vought	1	178	1,780	2,665	76	4,700
Brewster	-	-	136	599	-	735
Goodyear	-	-	377	2,108	1,522	4,007
Total	**1**	**178**	**2,293**	**5,372**	**1,598**	**9,442**

Source: *Official Munitions Production of the United States by Months, July 1, 1940-August 31, 1945*, War Production Board, Civilian Production Administration, 1947.

R-2800-8/-8W power output

Power condition	Power output [hp]	Revolutions [RPM]	Manifold pressure (boost) [mmHg]	Altitude [m]	Maximum duration [min.]
Takeoff	2,000	2,700	1,372	0	5
Emergency takeoff	1,850	2,800		0	1
Combat (WEP*)	2,250	2,700	1,461	0	5
	2,135	2,700	1,499	3,780	
	1,975	2,700	1,511	5,151	
Military	2,000	2,700	1,334	518	5
	1,800	2,700	1,346	4,785	
	1,650	2,700	1,346	6,400	
Normal rated (max. continuous)	1,675	2,550	1,118	1,676	no limits
	1,625	2,550	1,257	5,105	
	1,550	2,550	1,257	6,706	
Cruise max.	1,070	2,150	864	3,048	no limits
	970	2,150	864	6,248	
	950	2,050	864	7,925	

Notes: fuel grade 100/130; maximum permissible diving RPM: 3,060

* War Emergency Power – with water/methanol injection (only R-2800-8W)

Source: *Pilots Manual for F4U Corsair*, Aviation Publications 1989.

Specification – F4U-1/F3A-1/FG-1 Corsair

Model		XF4U-1	F4U-1 FG-1 F3A-1	F4U-1D FG-1D	F4U-2	XF4U-3A	FG-3	Corsair Mk IV
Dimensions:								
Wingspan	[m]	12.49	12.49	12.49	12.49	12.49	12.49	12.09
Wingspan (folded)	[m]	5.19	5.19	5.19	5.19	5.19	5.19	5.19
Length	[m]	9.73	10.16	10.16	10.16	10.24	10.24	10.16
Height*	[m]		4.48	4.48	4.48	4.51	4.51	4.48
Wing area	[m²]	29.17	29.17	29.17	29.17	29.17	29.17	28.34
Weights:								
Empty weight	[kg]	3,404	4,074	4,089	4,018	4,100	4,330	4,111
Takeoff weight	[kg]	4,244	5,461	5,461	5,192	5,272		5,492
Max. takeoff weight	[kg]	4,763	6,352	6,539	6,392	5,962	6,816	6,595
Internal tankage	[liter]	1,033	1,374	897	1,374	1,374	897	897
Auxiliary fuel tanks	[liter]	-	1×662**	1×662 or 2×568	1×606	-	2×568	2×623
Performance:								
Max. speed at a given altitude	[kph] [m]	652 2,926	671 6,096	671 6,096	616 7,681	663 9,144	755 10,241	668 5,944
Max. speed at sea level	[kph]		578	589	528	505	595	
Cruise speed	[kph]	298	293	293	301	290		420
Landing speed	[kph]	117	140	140	132	134	144	
Initial climb rate	[m/s]	13.5	14.7	17.1	15.1	15.2	21.4	16.0
Time to climb to 3,048/6,096 m	[min.]	~4.0/9.0	3.6/7.7	3.3/7.1	3.8/8.4	3.5/7.3	2.4/5.0	-/10.1
Service ceiling	[m]	9,449–10,800	11,247	12,192	10,272	11,704	13,929	10,698
Range	[km]	1,368	1,633	1,633	1,222	1,255	1,319	1,342
Max. range	[km]	1,722	2,570	3,186	2,993	2,301	3,459	2,513
Powerplant:								
Engine		XR-2800-2/-4	R-2800-8/-8W**	R-2800-8W	R-2800-8	XR-2800-16	R-2800-14W	R-2800-8W
Takeoff power	[hp]	1,850	2,000	2,000	2,000	2,000	2,100	2,000
Propeller		3-bladed	3-bladed	3-bladed	4-bladed	4-bladed	4-bladed	3-bladed
Propeller diameter	[m]	4.06	4.06	4.06/3.99	4.06	4.01	4.01	3.99
Armament:								
Guns: number × caliber (amount of ammunition)	[mm]	2×7.62 (1,500) 2×12.7 (600)	6×12.7 (2,350)	6×12.7 (2,400)***	5×12.7 (1,975)	6×12.7 (2,350)	6×12.7 (2,400)	6×12.7 (2,400)
Bomb load	[kg]	40×2.36	2×45.4/1×454**	2×454	2×45.4	2×45.4	2×726	2×454
Rockets (number × caliber)	[mm]	-	-	8×127 2×300	-	-	8×127 2×300	-

* On the ground, over propeller, three-point position; ** Only in F4U-1A/F3A-1A/FG-1A models; *** In F4U-1C model 4×20 mm (924).

Source: *Airplane Characteristics & Performance*, Bureau of Aeronautics; FAOW no. 88.

F4U-1 Corsair (BuNo unknown) coded 17-F-6 of VF-17, as it looked during carrier qualifications aboard USS Charger (CVE-30); the Atlantic, February 1943. VF-17 was formed in January 1943 at NAS Norfolk as the second (after VF-12) Corsair squadron of the US Navy. It entered combat in the Solomons in late October 1943, equipped with the newer F4U-1A model.

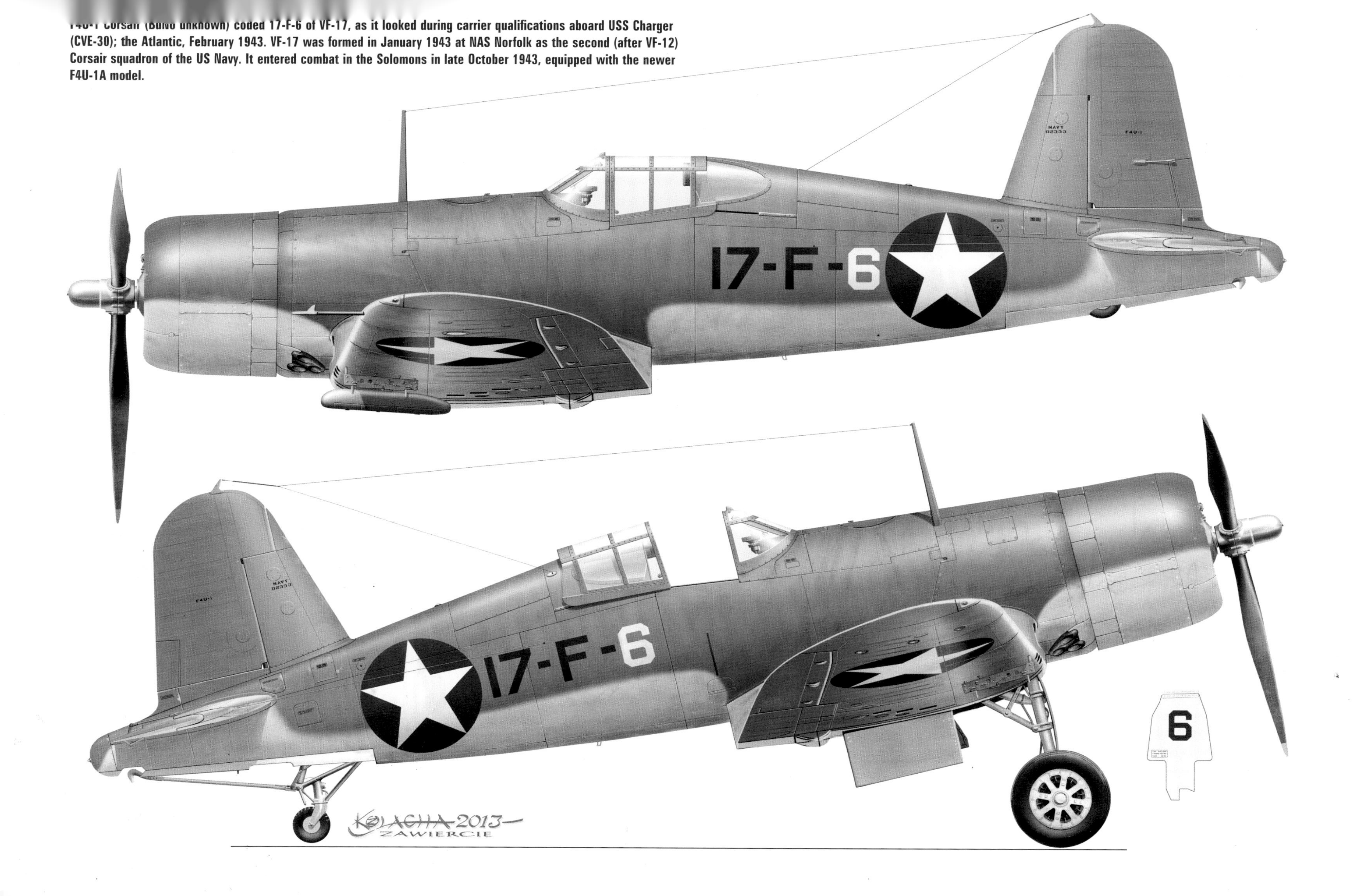

F4U-1A Corsair (BuNo 17656) No 5, flown by Lt(jg) Thomas Killefer (credited with 4.5 victories) of VF-17 *Jolly Rogers*; Bougainville, February 1944. On 5th March Killefer was forced to land this aircraft at Nissan, one of Green Islands, because of engine failure.

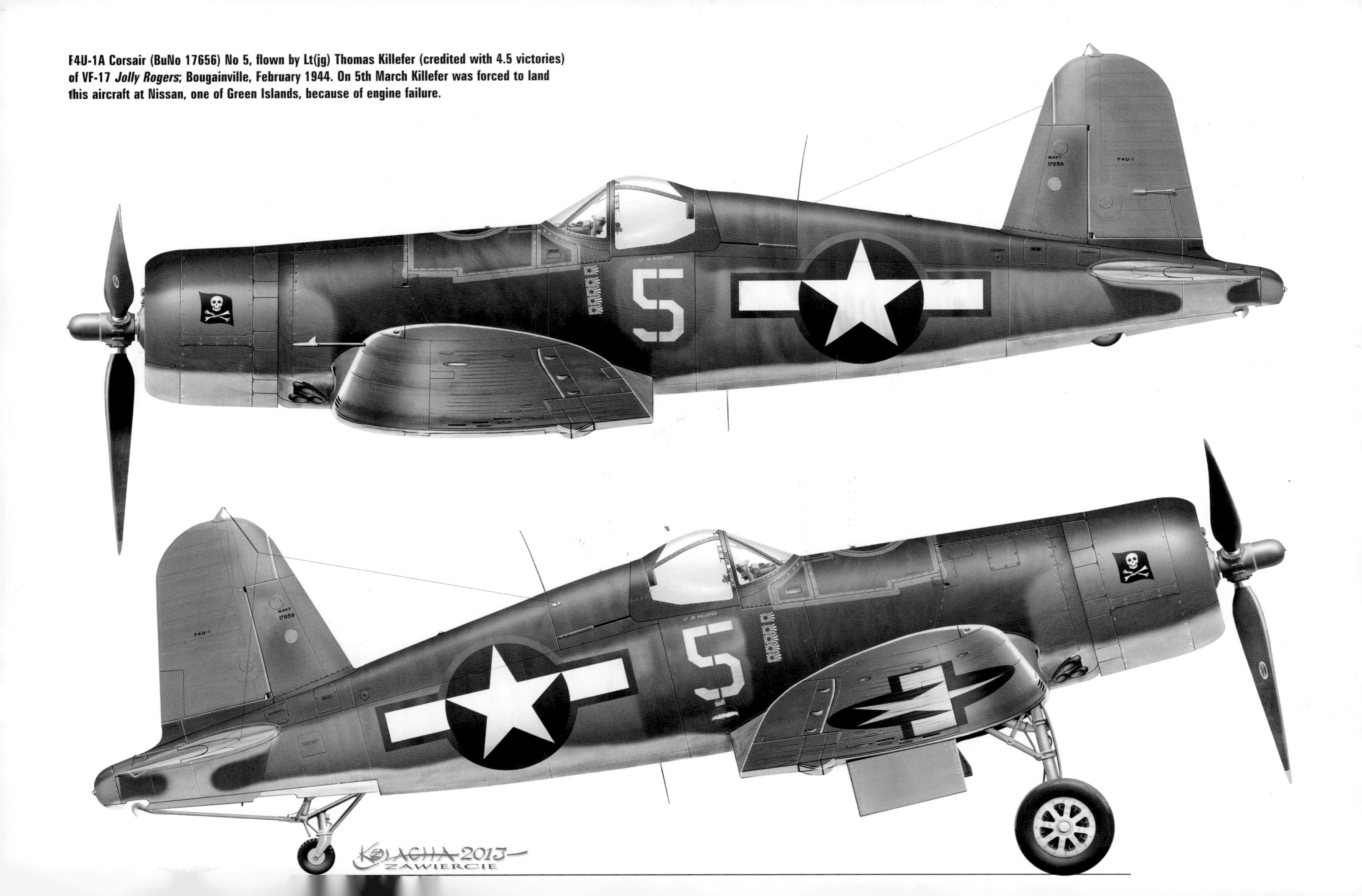

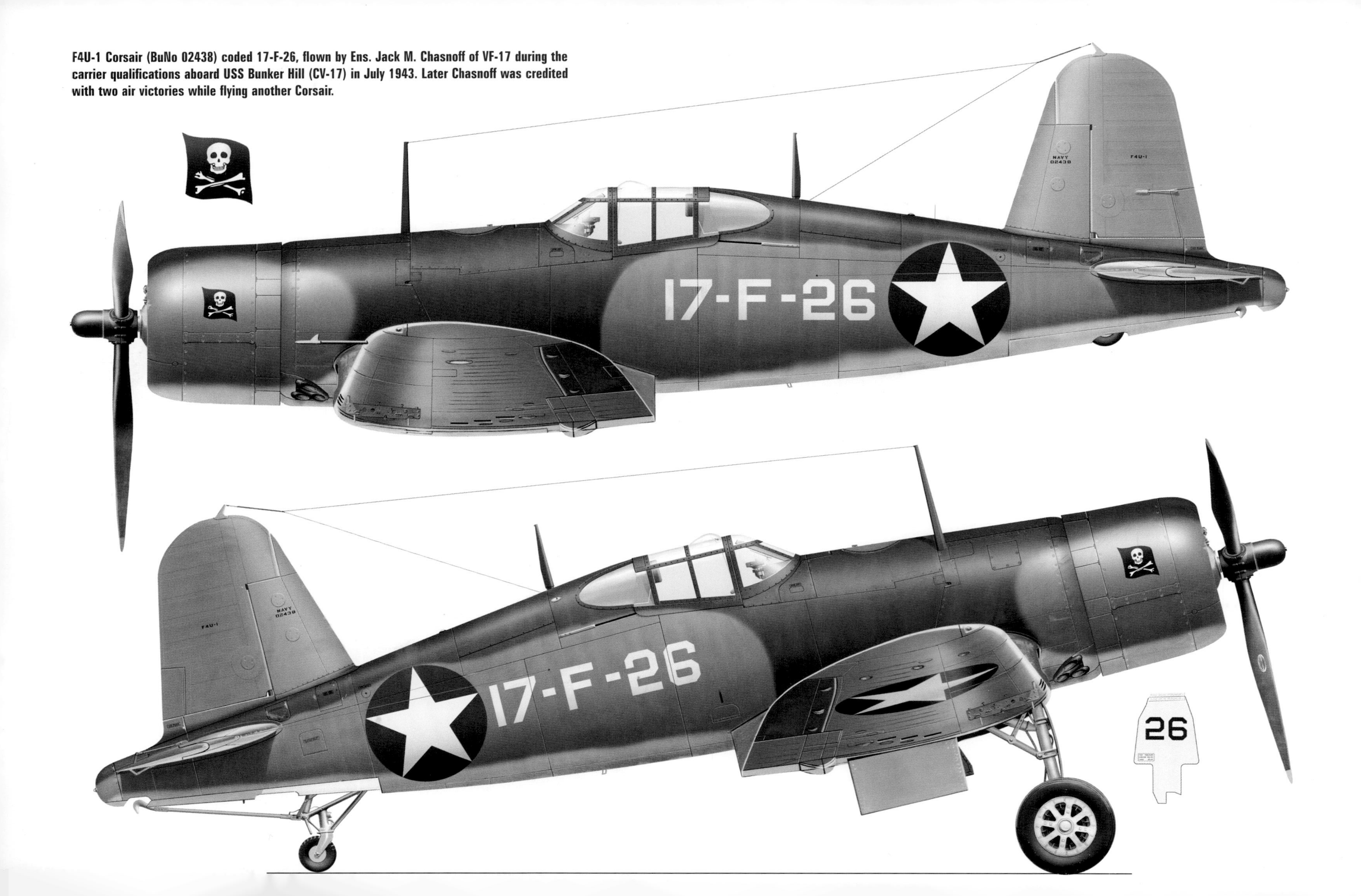

F4U-1 Corsair (BuNo 02438) coded 17-F-26, flown by Ens. Jack M. Chasnoff of VF-17 during the carrier qualifications aboard USS Bunker Hill (CV-17) in July 1943. Later Chasnoff was credited with two air victories while flying another Corsair.

F4U-1A Corsair (BuNo 17883) No 883 of VMF-214 *Black Sheep*, flown by, among others, Maj. Gregory Boyington and his wingman Lt. Robert W. McClurg; Barakoma, Vella Lavella, December 1943. Boyington scored 22 victories (and 4 probables) on Corsairs before he was shot down over Cape St. George on 3rd January 1944, ending up as a POW. McClurg tallied 7 victories (and two probables) by the war's end.

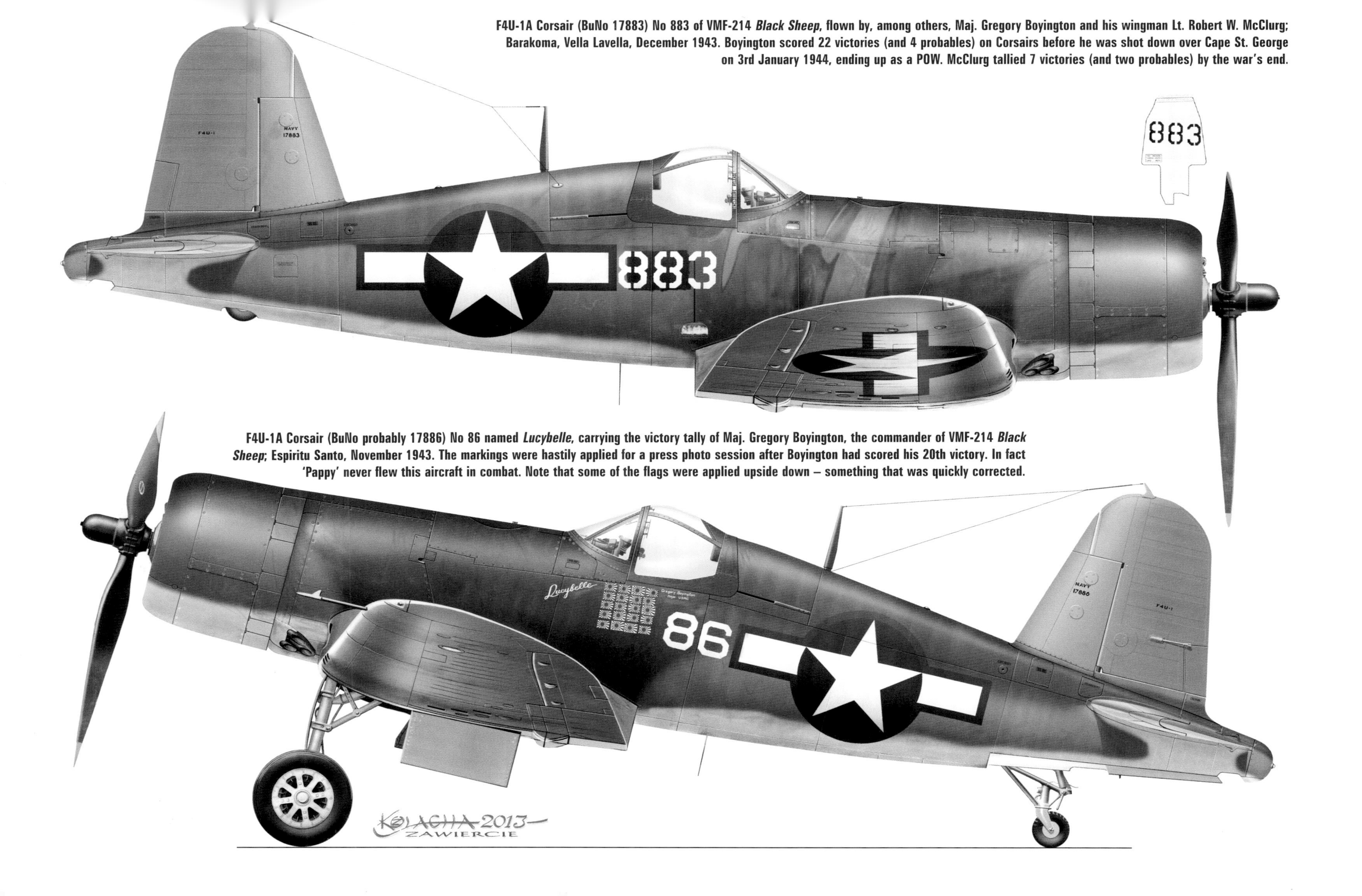

F4U-1A Corsair (BuNo probably 17886) No 86 named *Lucybelle*, carrying the victory tally of Maj. Gregory Boyington, the commander of VMF-214 *Black Sheep*; Espiritu Santo, November 1943. The markings were hastily applied for a press photo session after Boyington had scored his 20th victory. In fact 'Pappy' never flew this aircraft in combat. Note that some of the flags were applied upside down – something that was quickly corrected.

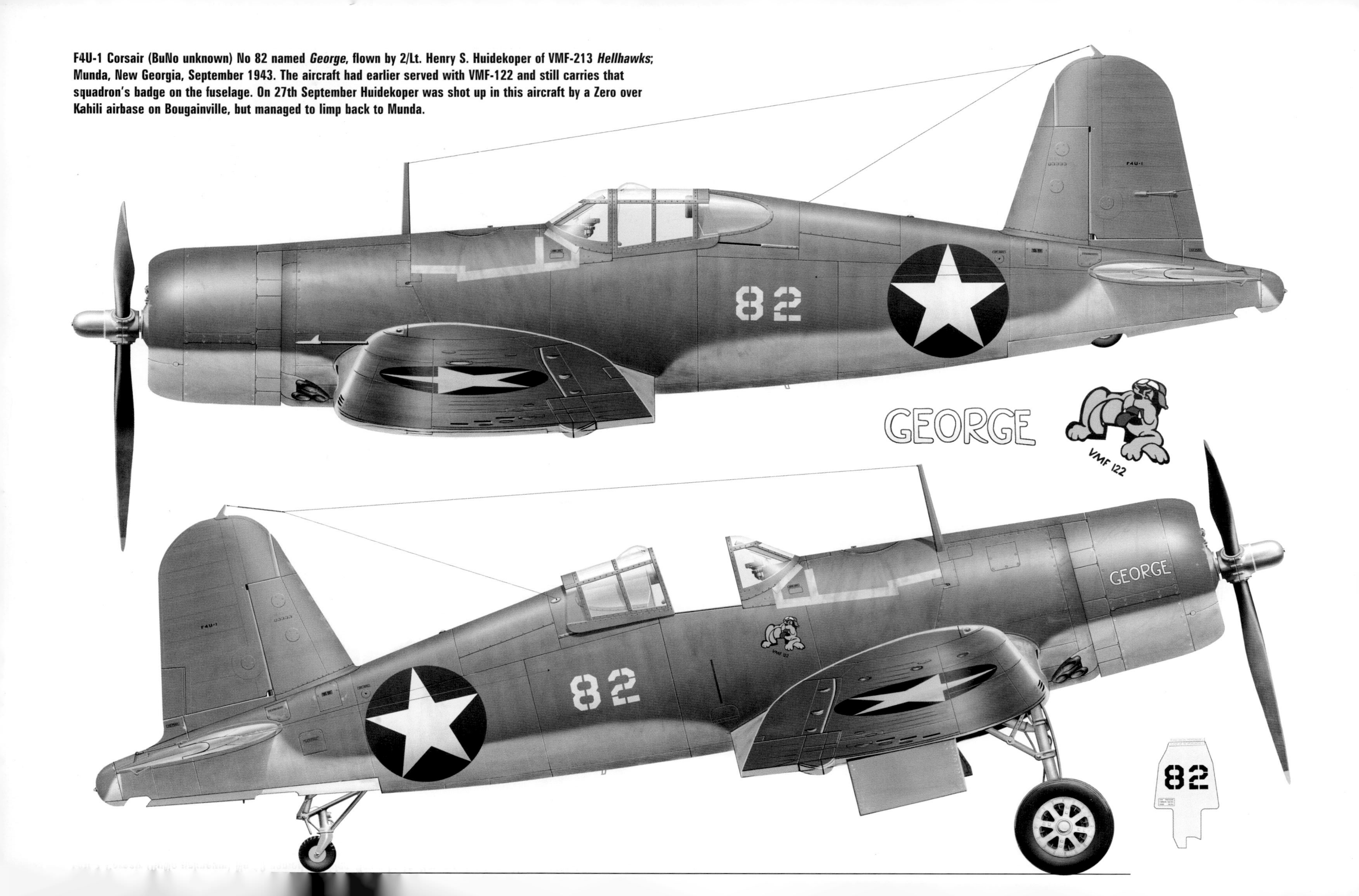

F4U-1 Corsair (BuNo unknown) No 82 named *George*, flown by 2/Lt. Henry S. Huidekoper of VMF-213 *Hellhawks*; Munda, New Georgia, September 1943. The aircraft had earlier served with VMF-122 and still carries that squadron's badge on the fuselage. On 27th September Huidekoper was shot up in this aircraft by a Zero over Kahili airbase on Bougainville, but managed to limp back to Munda.

F4U-1 Corsair (BuNo unknown) No 18 named *Bubbles*, flown by Lt. Howard J. 'Mick' Finn of VMF-124; Munda, New Georgia, summer 1943. During his first combat tour Finn scored five victories, including two, his fourth and fifth, on 15th August (two Val bombers near Vella Lavella), thus securing the ace status. Promoted to the rank of Captain in December 1943, he scored one more victory by the end of the war.

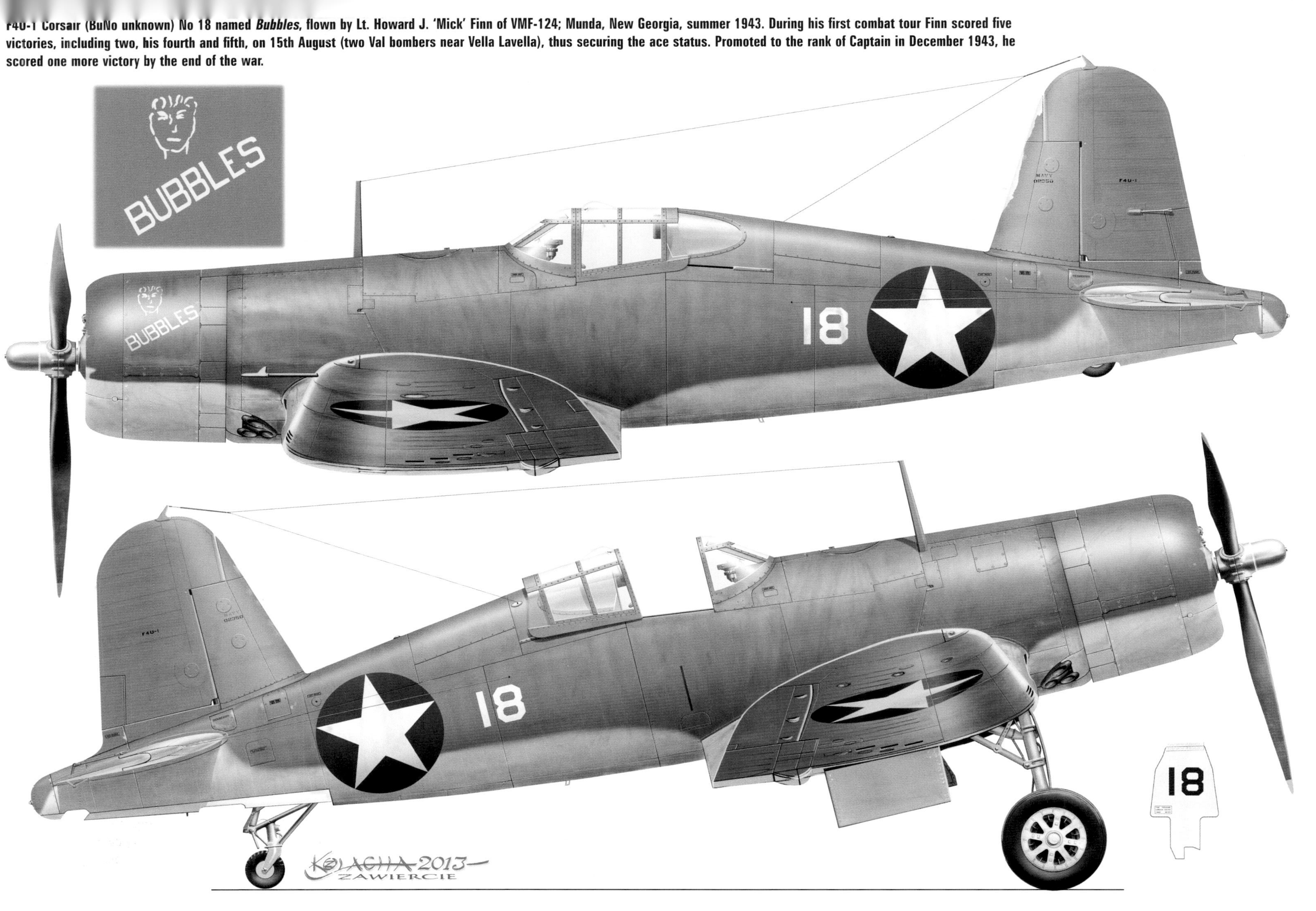

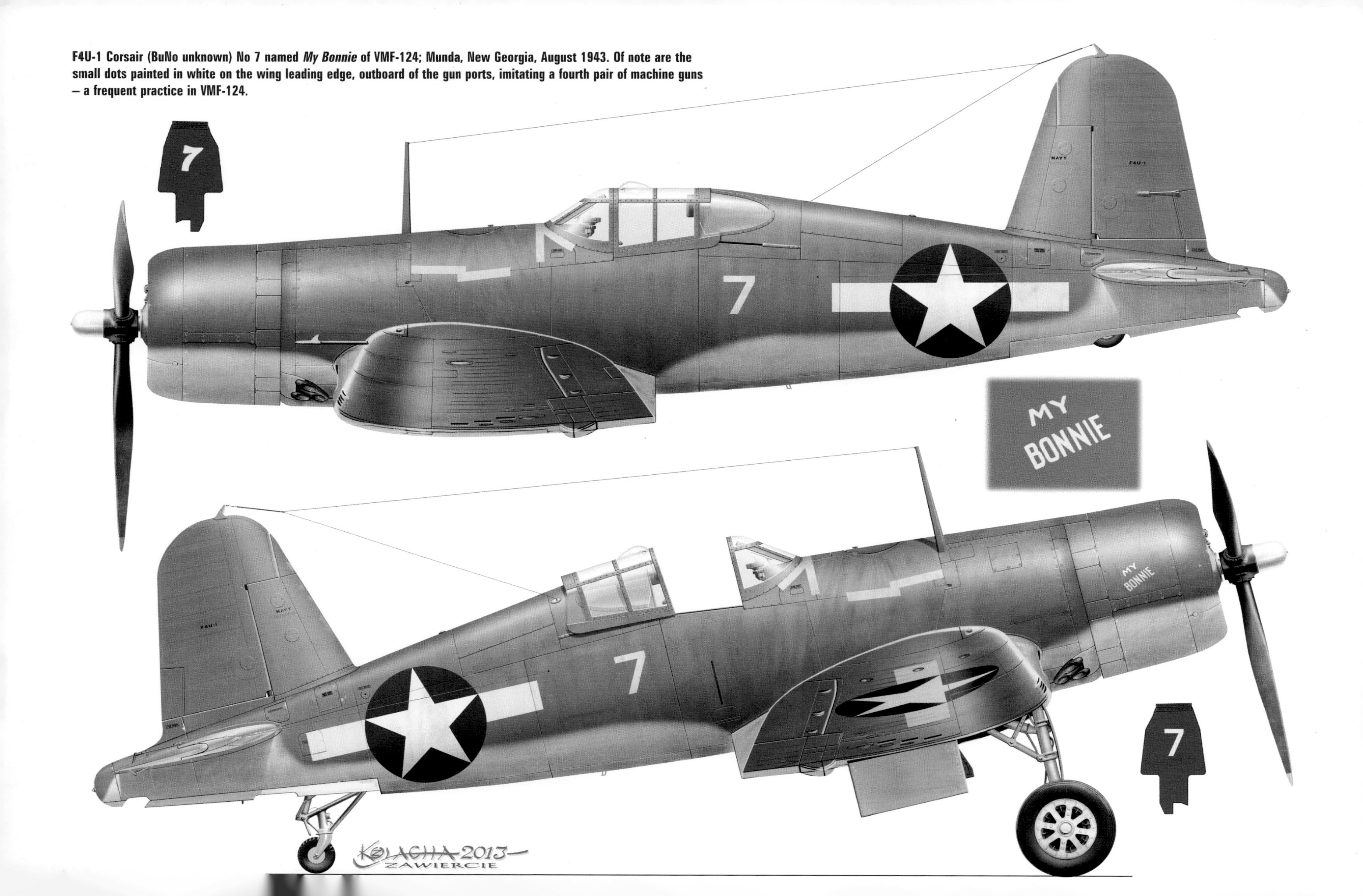

F4U-1 Corsair (BuNo unknown) No 7 named *My Bonnie* of VMF-124; Munda, New Georgia, August 1943. Of note are the small dots painted in white on the wing leading edge, outboard of the gun ports, imitating a fourth pair of machine guns – a frequent practice in VMF-124.